Lecture Notes in Physics

The Lecture Notes in Physics

The series Lecture Notes in Physics (LNP), founded in 1969, reports new developments in physics research and teaching – quickly and informally, but with a high quality and the explicit aim to summarize and communicate current knowledge in an accessible way. Books published in this series are conceived as bridging material between advanced graduate textbooks and the forefront of research to serve the following purposes:

• to be a compact and modern up-to-date source of reference on a well-defined topic;

• to serve as an accessible introduction to the field to postgraduate students and nonspecialist researchers from related areas;

• to be a source of advanced teaching material for specialized seminars, courses and schools.

Both monographs and multi-author volumes will be considered for publication. Edited volumes should, however, consist of a very limited number of contributions only. Proceedings will not be considered for LNP.

Volumes published in LNP are disseminated both in print and in electronic formats, the electronic archive is available at springerlink.com. The series content is indexed, abstracted and referenced by many abstracting and information services, bibliographic networks, subscription agencies, library networks, and consortia.

Proposals should be sent to a member of the Editorial Board, or directly to the managing editor at Springer:

Dr. Christian Caron
Springer Heidelberg
Physics Editorial Department I
Tiergartenstrasse 17
69121 Heidelberg/Germany
christian.caron@springer-sbm.com

W. Pötz   J. Fabian   U. Hohenester (Eds.)

Quantum Coherence

From Quarks to Solids

 Springer

Editors

Walter Pötz
Ulrich Hohenester
Institut für Theoretische Physik
Universität Graz
Universitätsplatz 5
8010 Graz, Austria
E-mail: walter.poetz@uni-graz.at
         ulrich.hohenester@uni-graz.at

Jaroslav Fabian
Institut für Theoretische Physik
Universität Regensburg
Universitätsstr. 31
93040 Regensburg, Germany
E-mail:
jaroslav.fabian@physik.uni-regensburg.de

W. Pötz, J. Fabian, U. Hohenester, *Quantum Coherence*,
Lect. Notes Phys. 689 (Springer, Berlin Heidelberg 2006), DOI 10.1007/b11398448

Library of Congress Control Number: 2005937152

ISSN 0075-8450
ISBN-10 3-540-30085-6 Springer Berlin Heidelberg New York
ISBN-13 978-3-540-30085-4 Springer Berlin Heidelberg New York

Springer is a part of Springer Science+Business Media
springer.com
© Springer-Verlag Berlin Heidelberg 2006
Printed in The Netherlands

Typesetting: by the authors and TechBooks using a Springer LaTeX macro package

Printed on acid-free paper   SPIN: 11398448   54/TechBooks   5 4 3 2 1 0

Preface

This volume is a collection of lecture notes which were provided by the key speakers of the Schladming Winter School in Theoretical Physics, "42. Internationale Universitätswochen für Theoretische Physik", in Schladming, Austria. This School took place from February 28 to March 6, 2004, and was given the title "Quantum Coherence in Matter: From Quarks to Solids."

Until 2003 the Schladming Winter School, which is organized by the Division for Theoretical Physics of the University of Graz, Austria, has been devoted primarily to topics in subatomic physics. A few years ago, however, it was decided to broaden the scope of this School and, in particular, to incorporate hot topics in condensed matter physics. This was done in an effort to better represent the scientific activities of the theory group of the Physics Department at the University of Graz.

The topic of the 42. School, "Quantum Coherence in Matter: From Quarks to Solids", was chosen for several reasons. Quantum coherence is a phenomenon that plays a crucial role in various forms of matter. The thriving field of quantum information, as one may say, the "hype" about quantum computers, and the open questions regarding the physical implementation of all these ideas, as well as conventional and unconventional approaches to use mesoscopic systems in future optoelectronic devices, make for a highly "hot topic" indeed. The compilation of lecture notes of the School is intended for advanced undergraduate and graduate students up to senior scientists who want to learn about or even get into this exciting field of physics. Research in this area is interdisciplinary and has fundamental, as well as applied aspects. Finally, because this was the first deviation from a school which hitherto has served purely particle physicists, we wanted to offer a topic which should be of general interest and with relevance to subatomic physics as well. Because the topic is very broad, we tried to provide examples from subatomic physics, atomic and molecular physics, as well as condensed matter physics. Also, we wanted to provide both fundamental issues, as well as more applied, implementation-oriented aspects.

Listed in alphabetical order, the speakers and topics of the School were R. Bertlmann (Vienna, Austria): "Entanglement, Decoherence and Bell Inequalities in Particle Physics"; M. Fleischhauer (Kaiserslautern, Germany): "Quantum Optics – Atomic Ensembles"; X. Hu (Buffalo, USA): "Quantum

Dot Quantum Computation"; S. Scheel (London, UK): "Quantum Gates and Decoherence"; T. Meier (Marburg, Germany): "Coherent Semiconductor Optics"; J. P. Paz (Buenos Aires, Argentina): "Decoherence and Ways to Combat It"; and W. H. Zurek (Los Alamos, USA): "Decoherence and Quantum Darwinism". One of the key speakers, C. Tejedor (Madrid), with "Exciton and Polariton Condensation" had to cancel his lectures on short notice, but kindly provided his lecture notes at the school, as did all other key speakers.

What has been said before about the flavor of the lectures also applies to the lecture notes presented in this volume. Of the seven key speakers who delivered their lectures, four provided manuscripts for publication. In addition, C. Tejedor's lecture notes are included.

R. Bertlmann from the University of Vienna, Austria, a former collaborator of J. S. Bell, discusses the basics of entanglement, decoherence, and Bell inequalities, as well as applications in and consequences to elementary particle physics. Moreover, he discusses the role of phases, such as the Berry and geometric phase.

S. Scheel and co-authors from the group of P. Knight at the Imperial College, London, UK, discuss quantum optical approaches to the realization of quantum gates. After an introduction to the basics of quantum gates, the authors discuss implementation based on a Mott transition in Bose condensates and a realization based on linear optics, and present a model for decoherence mechanisms in dielectric media.

X. Hu from the University of Buffalo, USA, presents the physics of quantum dots and their potential for implementation in a quantum computer. In the introduction he reviews qbit realizations based on the spin degree of freedom in quantum dots and Si-donors in GaAs. Electron spin coherence, spin manipulation, and the current experimental status are reviewed in the remainder of these lecture notes.

In "Coherent Semiconductor Optics", T. Meier and S. W. Koch from the University of Marburg, Germany, present a microscopic approach to the nonlinear optical properties of semiconductors within the density matrix approach at and beyond the Hartree–Fock level. Several truncation schemes are discussed. In particular, excitonic effects are studied and results are compared with experiment.

Finally, the lecture notes of C. Tejedor and co-workers deal with a review of theoretical aspects of exciton and polariton condensation. The main part of this work consists of a discussion of the standard theory of exciton condensation, magnetoexcitons, polariton condensates, and the polariton laser.

We are grateful to the lecturers for presenting their lectures in a very pedagogical way at the School, and for taking the time for preparing the manuscripts for publication in this book. We feel that this volume represents a good compilation on selected topics pertaining to quantum coherence phenomena in matter.

We acknowledge financial support from the main sponsors of the School, the Austrian Federal Ministry of Education, Science, and Culture, as well as from the Government of Styria. We have received financial, material, and technical assistance from the University of Graz, the town of Schladming, Ricoh Austria, and Hornig Graz. We also thank our colleagues at the Institute of Physics, our secretary, Ms. E. Monschein, and a number of undergraduate and graduate students for their valuable technical assistance, as well as all participants and speakers for making this School a success.

Graz
November 2005

Walter Pötz
Jaroslav Fabian
Ulrich Hohenester

Contents

Microscopic Theory of Coherent Semiconductor Optics

Exciton and Polariton Condensation

List of Contributors

Reinhold A. Bertlmann
Institut für Theoretische Physik
Boltzmanngasse 5
A-1090 Vienna, Austria
bertlman@ap.univie.ac.at

J. Fernandez-Rossier
Departamento de Fisica Aplicada
Universidad de Alicante
03080 Alicante
Spain
jfrossier@ua.es

Edward A. Hinds
Quantum Optics and Laser Science
Blackett Laboratory
Imperial College London
Prince Consort Road
London SW7 2BW
United Kingdom
ed.hinds@imperial.ac.uk

Xuedong Hu
Department of Physics
University at Buffalo
The State University
of New York
239 Fronczak Hall
Buffalo, NY 14260-1500
USA
xhu@buffalo.edu

Peter L. Knight
Quantum Optics and Laser Science
Blackett Laboratory
Imperial College London
Prince Consort Road
London SW7 2BW
United Kingdom
p.knight@imperial.ac.uk

Stephan W. Koch
Department of Physics and Material
Sciences Center
Philipps University
Renthof 5
D-35032 Marburg
Germany
stephan.w.koch@physik.
uni-marburg.de

Torsten Meier
Department of Physics and Material
Sciences Center
Philipps University
Renthof 5
D-35032 Marburg
Germany
torsten.meier@physik.
uni-marburg.de

Jiannis K. Pachos
DAMTP
Centre for Mathematical Sciences
Wilberforce Road
Cambridge CB3 0WA
United Kingdom
j.pachos@damtp.cam.ac.uk

Diego Porras
Max-Planck Institute for Quantum
Optics
D-85748 Garching, Germany
diego.porras@mpq.mpg.de

Stefan Scheel
Quantum Optics and Laser Science
Blackett Laboratory
Imperial College London
Prince Consort Road
London SW7 2BW
United Kingdom
s.scheel@imperial.ac.uk

Carlos Tejedor
Departamento de Fisica de la
Materia Condensada
Universidad Autonoma de Madrid
28049 Madrid
Spain
carlos.tejedor@uam.es

Entanglement, Bell Inequalities
and Decoherence in Particle Physics

R.A. Bertlmann

Institut für Theoretische Physik, Boltzmanngasse 5, A-1090 Vienna, Austria
Reinhold.Bertlmann@univie.ac.at

Abstract. We demonstrate the relevance of entanglement, Bell inequalities and decoherence in particle physics. In particular, we study in detail the features of the "strange" $K^0 \bar{K}^0$ system as an example of entangled meson–antimeson systems. The analogies and differences to entangled spin-$\frac{1}{2}$ or photon systems are worked out, the effects of a unitary time evolution of the meson system is demonstrated explicitly. After an introduction we present several types of Bell inequalities and show a remarkable connection to CP violation. We investigate the stability of entangled quantum systems pursuing the question of how possible decoherence might arise due to the interaction of the system with its "environment". The decoherence is strikingly connected to the entanglement loss of common entanglement measures. Finally, some outlook of the field is presented.

1 Introduction

In 1935, in his famous trilogy on "The present situation of quantum mechanics", Erwin Schrödinger [1] already realized the peculiar features of what he called entangled states – "... *verschränkte Zustände...*" was actually his German phrasing – in connection with quantum systems extended over physically distant parts. In the same year Einstein, Podolsky and Rosen (EPR) [2] constructed a gedanken experiment for a quantum system of two distant particles to demonstrate: *Quantum mechanics is incomplete*! Niels Bohr [3] replied immediately to the EPR article. His message was: *Quantum mechanics is complete*!

Also in 1935, W.H. Furry [4] emphasized, inspired by EPR and Schrödinger, the differences between the predictions of quantum mechanics (QM) of non-factorizable systems and models with spontaneous factorization.

However, the EPR–Bohr debate was regarded as rather philosophical and thus not very valuable for physicists for about 30 years until John Stewart Bell [5] brought this issue up again in his seminal work of 1964 "On the Einstein–Podolsky–Rosen Paradox", which caused a dramatic change in the view about this subject. Bell discovered what has since become known as *Bell's Theorem:*

- *No local hidden variable theory can reproduce all possible results of QM!*

R.A. Bertlmann: *Entanglement, Bell Inequalities and Decoherence in Particle Physics*,
Lect. Notes Phys. **689**, 1–45 (2006)
www.springerlink.com

It is achieved by establishing an inequality satisfied by the expectation values of all *local realistic theories* (LRT) but violated by the predictions of QM. Inequalities of this type are nowadays named quite generally *Bell inequalities* (BI).

It is the *nonlocality* arising from quantum entanglement – the *"spooky action at distance"* as we concluded in the 1980s [6] – which is the basic feature of quantum physics and is so contrary to our intuition.

Many beautiful experiments have been carried out over the years (see e.g. [7, 8, 9, 10]) by using the entanglement of the polarization of two photons; all (also the long-distance [11] or out-door experiments [12]) confirm impressively: *Nature contains a spooky action at distance!*

The nonlocality does not conflict with Einstein's relativity, so it cannot be used for superluminal communication, nevertheless, Bell's work [5, 13] initiated new physics, like quantum cryptography [11, 14, 15, 16] and quantum teleportation [17, 18], and it triggered a new technology: quantum information and quantum communication [19, 20]. More about "from Bell to quantum information" can be found in the book [21].

1.1 Particle Physics

Of course, it is of great interest to investigate the EPR–Bell correlations of measurements also for massive systems in particle physics. Here we want to work out the analogies and differences to the spin-$\frac{1}{2}$ or photon systems. Already in 1960, Lee and Yang [22] and several other authors [23, 24, 25] emphasized the EPR-like features of the "strange" $K^0\bar{K}^0$ system in a $J^{PC} = 1^-$ state, where the quantum number *strangeness* $S = +, -$ plays the role of spin $\Uparrow$ or $\Downarrow$. Indeed many authors [26, 27, 28, 29, 30, 31, 32, 33] suggested to investigate the $K^0\bar{K}^0$ pairs which are produced at the Φ resonance, for instance in the e^+e^--machine DAΦNE at Frascati. The nonseparability of the neutral kaon system – created in $p\bar{p}$–collisions – has been already analyzed by the authors of [34, 35, 36].

Similar systems are the entangled beauty mesons, $B^0\bar{B}^0$ pairs, produced at the $\Upsilon(4S)$ resonance (see e.g., [37, 38, 39, 40, 41, 42]), which we touch only little in this Article.

Specific realistic theories have been constructed [43, 44, 45, 46], which describe the $K^0\bar{K}^0$ pairs, as tests versus quantum mechanics. However, a general test of LRT versus QM relies on Bell inequalities, where – as we shall see – the different kaon detection times or the freely chosen kaon "quasi-spin" play the role of the different angles in the photon or spin-$\frac{1}{2}$ case. Furthermore, an interesting feature of kaons (and also of B-mesons) is CP violation (charge conjugation and parity) and indeed it turns out that BI imply bounds on the physical CP violation parameters.

The important difference of the kaon systems as compared to photons is their decay. We emphasize the necessity of including also the decay product states into the BI in order to have a unitary time evolution [47].

A main part of the article is also devoted to the investigation of the stability of the entangled quantum system. How possible decoherence might arise from the interaction of the system with its "environment", whatever this may be, and we will determine the strength of such effects. We study how decoherence is related to the loss of entanglement, and pursue the question: how to detect and quantify entanglement [48].

Of course, we cannot cover all subjects of the field. In Sect. 10 we mention interesting works we could not describe here and give an outlook to what we can expect in the near future.

Thus the contents of our article will be briefly the following:

- QM of K-mesons
- Introduction to BI
- BI for K-mesons, in time and "quasi-spin"
- Decoherence in entangled $K^0\bar{K}^0$ system
- Entanglement measures, entanglement loss and decoherence

2 QM of K-mesons

Neutral K-mesons are wonderful quantum systems! Four phenomena illustrate their "strange" behavior that is associated with the work of Abraham Pais [49, 50]:

(i)   *strangeness*
(ii) *strangeness oscillation*
(iii) *regeneration*
(iv) *CP violation.*

Let us start with a discussion of the properties of the neutral kaons, which we need in the following.

2.1 Strangeness

K-mesons are characterized by their *strangeness* quantum number S

$$\begin{aligned}
S|K^0\rangle &= +|K^0\rangle\,, \\
S|\bar{K}^0\rangle &= -|\bar{K}^0\rangle\,.
\end{aligned} \tag{1}$$

As the K-mesons are pseudoscalars their parity P is minus and charge conjugation C transforms particle K^0 and anti-particle $\bar{K}^0$ into each other so that we have for the combined transformation CP (in our choice of phases)

$$\begin{aligned}
CP|K^0\rangle &= -|\bar{K}^0\rangle\,, \\
CP|\bar{K}^0\rangle &= -|K^0\rangle\,.
\end{aligned} \tag{2}$$

It follows that the orthogonal linear combinations

$$|K_1^0\rangle = \frac{1}{\sqrt{2}}\{|K^0\rangle - |\bar{K}^0\rangle\}\,,$$
$$|K_2^0\rangle = \frac{1}{\sqrt{2}}\{|K^0\rangle + |\bar{K}^0\rangle\} \tag{3}$$

are eigenstates of CP, a quantum number conserved in strong interactions

$$CP|K_1^0\rangle = +|K_1^0\rangle\,,$$
$$CP|K_2^0\rangle = -|K_2^0\rangle\,. \tag{4}$$

2.2 CP Violation

Due to weak interactions, that do not conserve *strangeness* and are in addition *CP violating*, the kaons decay and the physical states, which differ slightly in mass, $\Delta m = m_L - m_S = 3.49 \times 10^{-6}$ eV, but immensely in their lifetimes and decay modes, are the short- and long-lived states

$$|K_S\rangle = \frac{1}{N}\{p|K^0\rangle - q|\bar{K}^0\rangle\}\,,$$
$$|K_L\rangle = \frac{1}{N}\{p|K^0\rangle + q|\bar{K}^0\rangle\}\,. \tag{5}$$

The weights $p = 1 + \varepsilon$, $q = 1 - \varepsilon$, with $N^2 = |p|^2 + |q|^2$ contain the complex *CP violating parameter* ε with $|\varepsilon| \approx 10^{-3}$. *CPT invariance* is assumed; thus the short- and long-lived states contain the same CP violating parameter $\varepsilon_S = \varepsilon_L = \varepsilon$. Then the CPT Theorem [51, 52, 53] implies that time reversal T is violated too.

The short-lived K-meson decays dominantly into $K_S \longrightarrow 2\pi$ with a width or lifetime $\Gamma_S^{-1} \sim \tau_S = 0.89 \times 10^{-10}$ s and the long-lived K-meson decays dominantly into $K_L \longrightarrow 3\pi$ with $\Gamma_L^{-1} \sim \tau_L = 5.17 \times 10^{-8}$ s. However, due to CP violation we observe a small amount $K_L \longrightarrow 2\pi$. To appreciate the importance of CP violation let us remind that the enormous disproportion of matter and antimatter in our universe is regarded as a consequence of CP violation that occurred immediately after the Big Bang.

2.3 Strangeness Oscillation

K_S, K_L are eigenstates of the non–Hermitian "effective mass" Hamiltonian

$$H = M - \frac{i}{2}\,\Gamma \tag{6}$$

satisfying

$$H\,|K_{S,L}\rangle = \lambda_{S,L}\,|K_{S,L}\rangle \tag{7}$$

with

$$\lambda_{S,L} = m_{S,L} - \frac{i}{2}\Gamma_{S,L} \,. \tag{8}$$

Both mesons K^0 and $\bar{K}^0$ have transitions to common states (due to CP violation) therefore they mix, that means they *oscillate* between K^0 and $\bar{K}^0$ before decaying. Since the decaying states evolve – according to the Wigner-Weisskopf approximation – exponentially in time

$$|K_{S,L}(t)\rangle = e^{-i\lambda_{S,L}t}|K_{S,L}\rangle \,, \tag{9}$$

the subsequent time evolution for K^0 and $\bar{K}^0$ is given by

$$|K^0(t)\rangle = g_+(t)|K^0\rangle + \frac{q}{p}g_-(t)|\bar{K}^0\rangle \,,$$
$$|\bar{K}^0(t)\rangle = \frac{p}{q}g_-(t)|K^0\rangle + g_+(t)|\bar{K}^0\rangle \tag{10}$$

with

$$g_\pm(t) = \frac{1}{2}\left[\pm e^{-i\lambda_S t} + e^{-i\lambda_L t}\right] \,. \tag{11}$$

Supposing that a K^0 beam is produced at $t = 0$, e.g. by strong interactions, then the probability for finding a K^0 or $\bar{K}^0$ in the beam is calculated to be

$$\left|\langle K^0|K^0(t)\rangle\right|^2 = \frac{1}{4}\left\{e^{-\Gamma_S t} + e^{-\Gamma_L t} + 2\,e^{-\Gamma t}\cos(\Delta m t)\right\} \,,$$
$$\left|\langle \bar{K}^0|K^0(t)\rangle\right|^2 = \frac{1}{4}\frac{|q|^2}{|p|^2}\left\{e^{-\Gamma_S t} + e^{-\Gamma_L t} - 2\,e^{-\Gamma t}\cos(\Delta m t)\right\} \,, \tag{12}$$

with $\Delta m = m_L - m_S$ and $\Gamma = \frac{1}{2}(\Gamma_L + \Gamma_S)$.

The K^0 beam oscillates with frequency $\Delta m/2\pi$, where $\Delta m\,\tau_S = 0.47$. The oscillation is clearly visible at times of the order of a few τ_S, before all K_S have died out leaving only the K_L in the beam. So in a beam which contains only K_0 mesons at the time $t = 0$ the $\bar{K}_0$ will occur far from the production source through its presence in the K_L meson with equal probability as the K_0 meson. A similar feature occurs when starting with a $\bar{K}^0$ beam.

2.4 Regeneration of K_S

In a K-meson beam, after a few centimeters, only the long-lived kaon state survives. But suppose we place a thin slab of matter into the K_L beam then the short–lived state K_S is regenerated because the K^0 and $\bar{K}^0$ components of the beam are scattered/absorbed differently in the matter.

3 Analogies and Quasi-Spin

A good *optical analogy* to the phenomenon of strangeness oscillation is the following situation. Let us take a crystal that absorbs the different polarization states of a photon differently, say H (horizontal) polarized light strongly but V (vertical) polarized light only weakly. Then if we shine R (right circular) polarized light through the crystal, after some distance there is a large probability for finding L (left circular) polarized light.

In comparison with spin-$\frac{1}{2}$ particles, or with photons having the polarization directions V and H, it is especially useful to work with the *"quasi-spin"* picture for kaons [47], originally introduced by Lee and Wu [54] and Lipkin [25]. The two states $|K^0\rangle$ and $|\bar{K}^0\rangle$ are regarded as the quasi-spin states up $|\Uparrow\rangle$ and down $|\Downarrow\rangle$ and the operators acting in this quasi-spin space are expressed by Pauli matrices. So the strangeness operator S can be identified with the Pauli matrix σ_3, the CP operator with $(-\sigma_1)$ and CP violation is proportional to σ_2. In fact, the Hamiltonian (6) can be written as

$$H = a \cdot \mathbf{1} + \boldsymbol{b} \cdot \boldsymbol{\sigma} , \tag{13}$$

with

$$b_1 = b\cos\alpha, \quad b_2 = b\sin\alpha, \quad b_3 = 0 ,$$

$$a = \frac{1}{2}(\lambda_L + \lambda_S), \quad b = \frac{1}{2}(\lambda_L - \lambda_S) \tag{14}$$

($b_3 = 0$ due to CPT invariance), and the phase α is related to the CP parameter ε by

$$e^{i\alpha} = \frac{1 - \varepsilon}{1 + \varepsilon} . \tag{15}$$

Summarizing, we have the following K-meson – spin-$\frac{1}{2}$ – photon analogy:

K-meson	spin-$\frac{1}{2}$	photon	
$|K^0\rangle$	$|\Uparrow\rangle_z$		$|V\rangle$
$|\bar{K}^0\rangle$	$|\Downarrow\rangle_z$		$|H\rangle$
$|K_S\rangle$	$|\Rightarrow\rangle_y$	$|L\rangle = \frac{1}{\sqrt{2}}(|V\rangle - i|H\rangle)$	
$|K_L\rangle$	$|\Leftarrow\rangle_y$	$|R\rangle = \frac{1}{\sqrt{2}}(|V\rangle + i|H\rangle)$	

Entangled States: Quite generally, we call a state *entangled* if it is not *separable*, i.e. not a convex combination of product states.

Now, what we are actually interested in are entangled states of $K^0\bar{K}^0$ pairs, in analogy to the entangled spin up and down pairs, or photon pairs. Such states are produced by e^+e^--machines through the reaction $e^+e^- \rightarrow \Phi \rightarrow K^0\bar{K}^0$, in particular at DA$\Phi$NE, Frascati, or they are produced in $p\bar{p}$-collisions, like, e.g., at LEAR, CERN. There, a $K^0\bar{K}^0$ pair is created in a $J^{PC} = 1^-$ quantum state and thus antisymmetric under C and P, and is described at the time $t = 0$ by the entangled state

$$|\psi(t=0)\rangle = \frac{1}{\sqrt{2}}\left\{|K^0\rangle_l \otimes |\bar{K}^0\rangle_r - |\bar{K}^0\rangle_l \otimes |K^0\rangle_r\right\} , \qquad (16)$$

which can be rewritten in the $K_S K_L$-basis

$$|\psi(t=0)\rangle = \frac{N_{SL}}{\sqrt{2}}\left\{|K_S\rangle_l \otimes |K_L\rangle_r - |K_L\rangle_l \otimes |K_S\rangle_r\right\} , \qquad (17)$$

with $N_{SL} = \frac{N^2}{2pq}$. The neutral kaons fly apart and will be detected on the left (l) and right (r) side of the source. Of course, during their propagation the $K^0 \bar{K}^0$ oscillate and the K_S, K_L states will decay. This is an important difference to the case of spin-$\frac{1}{2}$ particles or photons which are quite stable.

4 Time Evolution – Unitarity

Now let us discuss more closely the time evolution of the kaon states [55]. At any instant t the state $|K^0(t)\rangle$ decays to a specific final state $|f\rangle$ with a probability proportional to the absolute square of the transition matrix element. Because of the unitarity of the time evolution the norm of the total state must be conserved. This means that the decrease in the norm of the state $|K^0(t)\rangle$ must be compensated for by the increase in the norm of the final states.

So starting at $t = 0$ with a K^0 meson, the state we have to consider for a complete t-evolution is given by

$$|K^0\rangle \longrightarrow a(t)|K^0\rangle + b(t)|\bar{K}^0\rangle + \sum_f c_f(t)|f\rangle , \qquad (18)$$

$$a(t) = g_+(t) \qquad \text{and} \qquad b(t) = \frac{q}{p}g_-(t) . \qquad (19)$$

The functions $g_\pm(t)$ are defined in (11). Denoting the amplitudes of the decays of the $K^0, \bar{K}^0$ to a specific final state f by

$$\mathcal{A}(K^0 \longrightarrow f) \equiv A_f \qquad \text{and} \qquad \mathcal{A}(\bar{K}^0 \longrightarrow f) \equiv \bar{A}_f , \qquad (20)$$

we have

$$\frac{d}{dt}|c_f(t)|^2 = |a(t)A_f + b(t)\bar{A}_f|^2 , \qquad (21)$$

and for the probability of the decay $K_0 \to f$ at a certain time τ

$$P_{K^0 \longrightarrow f}(\tau) = \int_0^\tau \frac{d}{dt}|c_f(t)|^2 dt . \qquad (22)$$

Since the state $|K^0(t)\rangle$ evolves according to a Schrödinger equation with "effective mass" Hamiltonian (6) the decay amplitudes are related to the Γ matrix by

$$\Gamma_{11} = \sum_f |\mathcal{A}_f|^2, \quad \Gamma_{22} = \sum_f |\bar{\mathcal{A}}_f|^2, \quad \Gamma_{12} = \sum_f \mathcal{A}_f^* \bar{\mathcal{A}}_f . \tag{23}$$

These are the Bell–Steinberger unitarity relations [55]; they are a consequence of probability conservation, and play an important role.

For our purpose the following formalism generalized to arbitrary quasi-spin states is quite convenient [47, 56]. We describe a complete evolution of mass eigenstates by a unitary operator $U(t,0)$ whose effect can be written as

$$U(t,0)\,|K_{S,L}\rangle = e^{-i\lambda_{S,L}t}\,|K_{S,L}\rangle + |\Omega_{S,L}(t)\rangle , \tag{24}$$

where $|\Omega_{S,L}(t)\rangle$ denotes the state of all decay products. For the transition amplitudes of the decay product states we then have

$$\langle \Omega_S(t)|\Omega_S(t)\rangle = 1 - e^{-\Gamma_S t} , \tag{25}$$

$$\langle \Omega_L(t)|\Omega_L(t)\rangle = 1 - e^{-\Gamma_L t} , \tag{26}$$

$$\langle \Omega_L(t)|\Omega_S(t)\rangle = \langle K_L|K_S\rangle(1 - e^{i\Delta mt}e^{-\Gamma t}) , \tag{27}$$

$$\langle K_{S,L}|\Omega_S(t)\rangle = \langle K_{S,L}|\Omega_L(t)\rangle = 0 . \tag{28}$$

Mass eigenstates (5) are normalized but due to CP violation not orthogonal

$$\langle K_L|K_S\rangle = \frac{2Re\{\varepsilon\}}{1 + |\varepsilon|^2} =: \delta . \tag{29}$$

Now we consider entangled states of kaon pairs, and we start at time $t = 0$ from the entangled state given in the $K_S K_L$ basis choice (17)

$$|\psi(t=0)\rangle = \frac{N^2}{2\sqrt{2}pq}\{|K_S\rangle_l \otimes |K_L\rangle_r - |K_L\rangle_l \otimes |K_S\rangle_r\} . \tag{30}$$

Then we get the state at time t from (30) by applying the unitary operator

$$U(t,0) = U_l(t,0) \otimes U_r(t,0) , \tag{31}$$

where the operators $U_l(t,0)$ and $U_r(t,0)$ act on the subspace of the left and of the right mesons according to the time evolution (24).

What we are finally interested in are the quantum mechanical probabilities for detecting, or not detecting, a specific quasi-spin state on the left side $|k_n\rangle_l$ and on the right side $|k_n\rangle_r$ of the source. For that we need the projection operators $P_{l,r}(k_n)$ on the left, right quasi-spin states $|k_n\rangle_{l,r}$ together with the projection operators that act onto the orthogonal states $Q_{l,r}(k_n)$

$$P_l(k_n) = |k_n\rangle_{l\,l}\langle k_n| \quad \text{and} \quad P_r(k_n) = |k_n\rangle_{r\,r}\langle k_n| , \tag{32}$$

$$Q_l(k_n) = \mathbf{1} - P_l(k_n) \quad \text{and} \quad Q_r(k_n) = \mathbf{1} - P_r(k_n) . \tag{33}$$

So starting from the initial state (30) the unitary time evolution (31) determines the state at a time t_r

$$|\psi(t_r)\rangle = U(t_r,0)|\psi(t=0)\rangle = U_l(t_r,0) \otimes U_r(t_r,0)|\psi(t=0)\rangle . \quad (34)$$

If we now measure a certain quasi-spin k_m at t_r on the right side means that we project onto the state

$$|\tilde{\psi}(t_r)\rangle = P_r(k_m)|\psi(t_r)\rangle . \quad (35)$$

This state, which is now a one-particle state of the left-moving particle, evolves until t_l when we measure another k_n on the left side and we get

$$|\tilde{\psi}(t_l,t_r)\rangle = P_l(k_n)U_l(t_l,t_r)P_r(k_m)|\psi(t_r)\rangle . \quad (36)$$

The probability of the joint measurement is given by the squared norm of the state (36). It coincides (due to unitarity, composition laws and commutation properties of l,r-operators) with the state

$$|\psi(t_l,t_r)\rangle = P_l(k_n)P_r(k_m)U_l(t_l,0)U_r(t_r,0)|\psi(t=0)\rangle , \quad (37)$$

which corresponds to a factorization of the time into an eigentime t_l on the left side and into an eigentime t_r on the right side.

Then we calculate the quantum mechanical probability $P_{n,m}(Y,t_l;Y,t_r)$ for finding a k_n at t_l on the left side *and* a k_m at t_r on the right side and the probability $P_{n,m}(N,t_l;N,t_l)$ for finding *no* such kaons by the following norms; and similarly the probability $P_{n,m}(Y,t_l;N,t_r)$ when a k_n at t_l is detected on the left but *no* k_m at t_r on the right

$$P_{n,m}(Y,t_l;Y,t_r) = ||P_l(k_n)P_r(k_m)U_l(t_l,0)U_r(t_r,0)|\psi(t=0)\rangle||^2 , \quad (38)$$
$$P_{n,m}(N,t_l;N,t_r) = ||Q_l(k_n)Q_r(k_m)U_l(t_l,0)U_r(t_r,0)|\psi(t=0)\rangle||^2 , \quad (39)$$
$$P_{n,m}(Y,t_l;N,t_r) = ||P_l(k_n)Q_r(k_m)U_l(t_l,0)U_r(t_r,0)|\psi(t=0)\rangle||^2 . \quad (40)$$

5 Bell Inequalities for Spin-$\frac{1}{2}$ Particles

In this section we derive the well-known Bell-inequalities [13] and we want to present the details because of the close analogy between the spin/photon systems and kaon systems. Let us start with a BI which holds most generally, the CHSH inequality, named after Clauser, Horne, Shimony and Holt [57], and then we derive from that inequality – with two further assumptions – the original Bell inequality and the Wigner-type inequality.

Let $A(n,\lambda)$ and $B(m,\lambda)$ be the definite values of two quantum observables $A^{QM}(n)$ and $B^{QM}(m)$, measured by *Alice* on one side and by *Bob* on the other. The parameter λ denotes the hidden variables which are not accessible to an experimenter but carry the additional information needed in a LRT. The measurement result of one observable is $A(n,\lambda) = \pm 1$ corresponding to the spin measurement "spin up" and "spin down" along the quantization direction n of particle 1; and $A(n,\lambda) = 0$ if no particle was detected at all. The analogue holds for the result $B(m,\lambda)$ of particle 2.

Bell's Locality Hypothesis: The basic ingredient is the following.

- *The outcome of Alice's measurement does not depend on the settings of Bob's instruments; i.e., $A(n, \lambda)$ depends only on the direction n, but not on m; and analogously $B(m, \lambda)$ depends only on m but not on n!*

That's the crucial point, for the combined spin measurement we then have the following expectation value

$$E(n, m) = \int d\lambda \ \rho(\lambda) A(n, \lambda) B(m, \lambda) , \qquad (41)$$

with the normalized probability distribution

$$\int d\lambda \ \rho(\lambda) = 1 . \qquad (42)$$

This quantity $E(n, m)$ corresponds to the quantum mechanical expectation value $E^{QM}(n, m) = \langle A^{QM}(n) \otimes B^{QM}(m) \rangle$.

It is straightforward to estimate the absolute value of the difference of two expectation values (see, for example, [57, 58, 59]):

$$E(n, m) - E(n, m') = \int d\lambda \ \rho(\lambda) A(n, \lambda) B(m, \lambda) \left\{ 1 \pm A(n', \lambda) B(m', \lambda) \right\}$$
$$- \int d\lambda \ \rho(\lambda) A(n, \lambda) B(m', \lambda) \left\{ 1 \pm A(n', \lambda) B(m, \lambda) \right\} , \qquad (43)$$

then the absolute value provides

$$| E(n, m) - E(n, m') | \ \leq \int d\lambda \ \rho(\lambda) \left\{ 1 \pm A(n', \lambda) B(m', \lambda) \right\}$$
$$+ \int d\lambda \ \rho(\lambda) \left\{ 1 \pm A(n', \lambda) B(m, \lambda) \right\} , \qquad (44)$$

and with the normalization (42) we get

$$| E(n, m) - E(n, m') | \ \leq 2 \pm | E(n', m') + E(n', m) | , \qquad (45)$$

or written more symmetrically

$$S = | E(n, m) - E(n, m') | + | E(n', m') + E(n', m) | \ \leq \ 2 . \qquad (46)$$

This is the familiar *CHSH-inequality*, derived by Clauser, Horne, Shimony and Holt [57] in 1969. *Every local realistic hidden variable theory must obey this inequality!*

Calculating the quantum mechanical expectation values $E^{QM}(n, m)$ in the spin singlet state $|\psi\rangle = \frac{1}{\sqrt{2}} \left(|\Uparrow_n\rangle |\Downarrow_m\rangle - |\Downarrow_n\rangle |\Uparrow_m\rangle \right)$

$$E^{QM}(n, m) = \langle \psi | A^{QM}(n) \otimes B^{QM}(m) | \psi \rangle$$
$$= \langle \psi | \boldsymbol{\sigma} \cdot \boldsymbol{n} \otimes \boldsymbol{\sigma} \cdot \boldsymbol{m} | \psi \rangle = - \cos \phi_{n,m} , \qquad (47)$$

where the $\phi_{n,m}$ are the angles between the two quantization directions n and m, we insert (47) into (46) and obtain the following inequality

$$S(n, m, n', m') = |\cos \phi_{n,m} - \cos \phi_{n,m'}|$$
$$+ |\cos \phi_{n',m'} + \cos \phi_{n',m}| \leq 2. \qquad (48)$$

For certain angles ϕ – the so called *Bell angles* – inequality (48) is violated! The maximal violation is $2\sqrt{2}$, achieved by the Bell angles $\phi_{n,m'} = \frac{3\pi}{4}$ and $\phi_{n,m} = \phi_{n',m'} = \phi_{n',m} = \frac{\pi}{4}$.

Experimentally, for entangled photon pairs, inequality (48) is violated under strict Einstein locality conditions in an impressive way, with a result in close agreement with QM $S_{\exp} = 2.73 \pm 0.02$ [10], confirming previous experimental results on similar inequalities [7, 8, 9].

Now we make two assumptions: perfect correlations $E(n, n) = -1$, no 0 results, and choose 3 different angles (e.g. $n' = m'$) then inequality (45) gives

$$|E(n, m) - E(n, n')| \leq 2 \pm \underbrace{\{E(n', n')}_{-1 \forall n'} + E(n', m)\}$$

or

$$|E(n, m) - E(n, n')| \leq 1 + E(n', m). \qquad (49)$$

This is the famous *Bell's inequality* derived by J.S. Bell in 1964 [5].

Finally, we rewrite the expectation value for the measurement of two spin-$\frac{1}{2}$ particles in terms of probabilities P

$$E(n, m) = P(n \Uparrow; m \Uparrow) + P(n \Downarrow; m \Downarrow) - P(n \Uparrow; m \Downarrow) - P(n \Downarrow; m \Uparrow)$$
$$= -1 + 4 P(n \Uparrow; m \Uparrow), \qquad (50)$$

where we used $P(n \Uparrow; m \Uparrow) = P(n \Downarrow; m \Downarrow)$, $P(n \Uparrow; m \Downarrow) = P(n \Downarrow; m \Uparrow)$ and $\sum P = 1$. Then Bell's original inequality (49) turns into *Wigner's inequality*

$$P(n; m) \leq P(n; n') + P(n'; m), \qquad (51)$$

where the P's are the probabilities for finding the spins up–up on the two sides or down-down or twisted, up-down and down-up. Note, that the Wigner inequality has been originally derived by a set-theoretical approach [60].

6 Bell Inequalities for K-mesons

Let us return to the kaon states which we describe within the "quasi-spin" picture. We start again from the state $|\psi(t = 0)\rangle$, (16) or (17), of entangled "quasi-spins" states and consider its time evolution $U(t, 0)|\psi(0)\rangle$. Then we find the following situation.

6.1 Analogies and Differences

When performing two measurements to detect the kaons at the same time at the left side and at the right side of the source, the probability of finding two mesons with the same "quasi-spin" – i.e. $K^0 K^0$ with strangeness $(+1, +1)$ or $\bar{K}^0 \bar{K}^0$ with strangeness $(-1, -1)$ – is zero.

That means, if we measure at time t a K^0 meson on the left side (denoted by Y, yes), we will find with certainty at the same time t *no* K^0 on the right side (denoted by N, no). This is an EPR–Bell correlation analogously to the spin-$\frac{1}{2}$ or photon case, e.g., with polarization V–H (see [30, 47, 61]).

The analogy would be perfect, if the kaons were stable ($\Gamma_S = \Gamma_L = 0$); then the quantum probabilities yield the result

$$P(Y, t_l; Y, t_r) = P(N, t_l; N, t_r) = \frac{1}{4}\{1 - \cos(\Delta m(t_l - t_r))\} ,$$

$$P(Y, t_l; N, t_r) = P(N, t_l; Y, t_r) = \frac{1}{4}\{1 + \cos(\Delta m(t_l - t_r))\} . \tag{52}$$

It coincides with the probability result of finding simultaneously two entangled spin-$\frac{1}{2}$ particles in spin directions $\Uparrow \Uparrow$ or $\Uparrow \Downarrow$ along two chosen directions $\boldsymbol{n}$ and $\boldsymbol{m}$

$$P(\boldsymbol{n}, \Uparrow; \boldsymbol{m}, \Uparrow) = P(\boldsymbol{n}, \Downarrow; \boldsymbol{m}, \Downarrow) = \frac{1}{4}\{1 - \cos\theta\} ,$$

$$P(\boldsymbol{n}, \Uparrow; \boldsymbol{m}, \Downarrow) = P(\boldsymbol{n}, \Downarrow; \boldsymbol{m}, \Uparrow) = \frac{1}{4}\{1 + \cos\theta\} . \tag{53}$$

Analogies: Perfect analogy between times and angles.

- *The time differences $\Delta m(t_l - t_r)$ in the kaon case play the role of the angle differences θ in the spin-$\frac{1}{2}$ or photon case.*

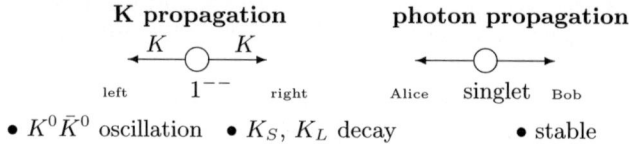

- for $t_l = t_r$: EPR-like correlation

- for $t_l \neq t_r$: EPR–Bell correlation

Differences: There are important physical differences.

(i) While in the spin-$\frac{1}{2}$ or photon case one can test whether a system is in an arbitrary spin state $\alpha| \Uparrow\rangle + \beta| \Downarrow\rangle$ one cannot test it for an arbitrary superposition $\alpha|K^0\rangle + \beta|\bar{K}^0\rangle$.

(ii) For entangled spin-$\frac{1}{2}$ particles or photons it is clearly sufficient to consider the direct product space $H^l_{spin} \otimes H^r_{spin}$ to account for all spin or polarization properties of the entangled system, however, this is not so for kaons. The unitary time evolution of a kaon state also involves the decay product states (see Sect. 4), therefore one has to include the Hilbert space of the decay products $H^l_\Omega \otimes H^r_\Omega$ which is orthogonal to the space $H^l_{kaon} \otimes H^r_{kaon}$ of the surviving kaons.

Consequently, the appropriate dichotomic question on the system is: "*Are you a K^0 or not?*" It is clearly different from the question "*Are you a K^0 or a $\bar{K}^0$?*" (as treated, e.g., in [61]), since all decay products – an additional characteristic of the quantum system – are ignored by the latter.

6.2 Bell-CHSH Inequality – General Form

Measuring a $\bar{K}^0$ (it is the antiparticle that is actually measured via strong interactions in matter) on the left side we can predict with certainty to find at the same time *no* $\bar{K}^0$ at the right side. In any LRT this property *no* $\bar{K}^0$ must be present on the right side irrespective of having the measurement performed or not. In order to discriminate between QM and LRT we set up a Bell inequality for the kaon system where now the different times play the role of the different angles in the spin-$\frac{1}{2}$ or photon case. But, in addition, we also may use the freedom of choosing a particular quasi-spin state of the kaon, e.g., the strangeness eigenstate, the mass eigenstate, or the CP eigenstate. Thus an expectation value for the combined measurement $E(k_n, t_a; k_m, t_b)$ depends on a certain quasi-spin k_n measured on the left side at a time t_a and on a (possibly different) k_m on the right side at t_b. Taking over the argumentation of Sect. 5 we derive the following *Bell-CHSH inequality* [47]

$$|E(k_n, t_a; k_m, t_b) - E(k_n, t_a; k_{m'}, t_{b'})|$$
$$+ |E(k_{n'}, t_{a'}; k_{m'}, t_{b'}) + E(k_{n'}, t_{a'}; k_m, t_b)| \leq 2, \quad (54)$$

which expresses both the freedom of choice in time *and* in quasi-spin. If we identify $E(k_n, t_a; k_m, t_b) \equiv E(n, m)$ we are back at the inequality (46) for the spin-$\frac{1}{2}$ case.

The expectation value for the series of identical measurements can be expressed in terms of the probabilities, where we denote by $P_{n,m}(Y, t_a; Y, t_b)$ the probability for finding a k_n at t_a on the left side and finding a k_m at t_b on the right side and by $P_{n,m}(N, t_a; N, t_b)$ the probability for finding *no* such kaons; similarly $P_{n,m}(Y, t_a; N, t_b)$ denotes the case when a k_n at t_a is detected on the left but *no* k_m at t_b on the right. Then the expectation value is given by the following probabilities

$$E(k_n, t_a; k_m, t_b) = P_{n,m}(Y, t_a; Y, t_b) + P_{n,m}(N, t_a; N, t_b)$$
$$-P_{n,m}(Y, t_a; N, t_b) - P_{n,m}(N, t_a; Y, t_b). \qquad (55)$$

Since the sum of the probabilities for (Y, Y), (N, N), (Y, N) and (N, Y) just add up to unity we get

$$E(k_n, t_a; k_m, t_b) = -1 + 2\left\{P_{n,m}(Y, t_a; Y, t_b) + P_{n,m}(N, t_a; N, t_b)\right\}. \quad (56)$$

Note that relation (55) between the expectation value and the probabilities is satisfied for QM and LRT as well.

6.3 Bell Inequality for Time Variation

Alternative: In Bell inequalities for meson systems we have an option

- fixing the quasi-spin – freedom in time
- freedom in quasi-spin – fixing time.

Let us elaborate on the first one. We choose a definite quasi-spin, say strangeness $S = +1$ that means $k_n = k_m = k_{n'} = k_{m'} = K^0$, we neglect CP violation (which does not play a role to our accuracy level here) then we obtain the following formula for the expectation value

$$E(t_l; t_r) = -\cos(\Delta m \Delta t) \cdot e^{-\Gamma(t_l + t_r)}$$
$$+\frac{1}{2}(1 - e^{-\Gamma_L t_l})(1 - e^{-\Gamma_S t_r}) + \frac{1}{2}(1 - e^{-\Gamma_S t_l})(1 - e^{-\Gamma_L t_r}). \ (57)$$

Since expectation value (57) corresponds to a unitary time evolution, it contains, in addition to the pure meson state contribution, terms coming from the decay product states $|\Omega_{L,S}(t)\rangle$.

However, in the kaon system we can neglect the width of the long-lived K-meson as compared to the short-lived one, $\Gamma_L \ll \Gamma_S$, so that we have to a good approximation

$$E^{\text{approx}}(t_l; t_r) = -\cos(\Delta m \Delta t) \cdot e^{-\Gamma(t_l + t_r)}, \qquad (58)$$

which coincides with an expectation value where all decay products are ignored (the probabilities, e.g. $P(K^0, t_a; \bar{K}^0, t_b)$, just contain the meson states). This is certainly not the case for other meson systems, like the $B^0\bar{B}^0$, $D^0\bar{D}^0$ and $B_s^0\bar{B}_s^0$ systems (see below).

Inserting now the quantum mechanical expectation value (58) into inequality (54) we arrive at Ghirardi, Grassi and Weber's result [56]

$$|e^{-\frac{\Gamma_S}{2}(t_a + t_{a'})}\cos(\Delta m(t_a - t_{a'})) - e^{-\frac{\Gamma_S}{2}(t_a + t_{b'})}\cos(\Delta m(t_a - t_{b'}))| \qquad (59)$$
$$+|e^{-\frac{\Gamma_S}{2}(t_{a'} + t_b)}\cos(\Delta m(t_{a'} - t_b)) + e^{-\frac{\Gamma_S}{2}(t_b + t_{b'})}\cos(\Delta m(t_b - t_{b'}))| \leq 2.$$

(Of course, we could have chosen $\bar{K}^0$ instead of K^0 without any change).

No Violation: Unfortunately, inequality (59) *cannot be violated* [56, 62] for any choice of the four (positive) times $t_a, t_b, t_{a'}, t_{b'}$ due to the interplay between the kaon decay and strangeness oscillations. As demonstrated in [63] a possible violation depends very much on the ratio $x = \Delta m/\Gamma$. The numerically determined range for *no violation* is $0 < x < 2$ [64] and the experimental value $x_{exper} = 0.95$ lies precisely inside.

Remark on Other Meson Systems: Instead of K-mesons we also can consider entangled B–mesons, produced via the resonance decay $\Upsilon(4S) \rightarrow B^0 \bar{B}^0$, e.g., at the KEKB asymmetric e^+e^- collider in Japan. In such a system, the *beauty* quantum number $B = +, -$ is the analogue to *strangeness* $S = +, -$ and instead of long- and short-lived states we have the heavy $|B_H\rangle$ and light $|B_L\rangle$ as eigenstates of the non–Hermitian "effective mass" Hamiltonian. Since for B-mesons the decay widths are equal, $\Gamma_H = \Gamma_L = \Gamma_B$, we get for the expectation value in a unitary time evolution

$$E(t_l; , t_r) = -\cos(\Delta m_B \Delta t) \cdot e^{-\Gamma_B(t_l + t_r)} \\ + (1 - e^{-\Gamma_B t_l})(1 - e^{-\Gamma_B t_r}), \tag{60}$$

where $\Delta m_B = m_H - m_L$ is the mass difference of the heavy and light B-meson. Here, the additional term from the decay products cannot be ignored.

Inserting expectation value (60) into inequality (54) for a fixed quasi-spin, say, for flavor $B = +1$, i.e. B^0, we find that the Bell-CHSH inequality *cannot be violated* in the x range $0 < x < 2.6$ [64]. Again, the experimental value $x_{exper} = 0.77$ lies inside.

Precisely the same feature occurs for an other meson–antimeson system, the *charmed* system $D^0 \bar{D}^0$.

Since the experimental x values for different meson systems are the following ones:

x	Meson System
0.95	$K^0 \bar{K}^0$
0.77	$B^0 \bar{B}^0$
< 0.03	$D^0 \bar{D}^0$
> 19.00	$B_s^0 \bar{B}_s^0$

no violation of the Bell–CHSH inequality occurs for the familiar meson-antimeson systems; only for the last system a violation is expected.

Conclusion: One cannot use the time-variation type of Bell inequality to exclude local realistic theories.

6.4 Bell Inequality for Quasi-Spin States – CP Violation

Now we investigate the second option. We fix the time, say at $t = 0$, and vary the quasi-spin of the K-meson. It corresponds to a rotation in quasi-spin space analogously to the spin-$\frac{1}{2}$ or photon case.

Analogy: Rotation in "quasi-spin" space $\longleftrightarrow$ polarization space

- *The quasi-spin of kaons plays the role of spin or photon polarization!*

$$|k\rangle = a|K^0\rangle + b|\bar{K}^0\rangle \quad \longleftrightarrow \quad |n\rangle = \cos\frac{\alpha}{2}|\Uparrow\rangle + \sin\frac{\alpha}{2}|\Downarrow\rangle$$

For a BI we need 3 different "angles" – "quasi-spins" and we may choose the H, S and CP eigenstates

$$|k_n\rangle = |K_S\rangle, \qquad |k_m\rangle = |\bar{K}^0\rangle, \qquad |k_{n'}\rangle = |K_1^0\rangle. \tag{61}$$

Denoting the probability of measuring the short-lived state K_S on the left side and the anti-kaon $\bar{K}^0$ on the right side, at the time $t = 0$, by $P(K_S, \bar{K}^0)$, and analogously the probabilities $P(K_S, K_1^0)$ and $P(K_1^0, \bar{K}^0)$ we can easily derive under the usual hypothesis of Bell's locality the following *Wigner–like Bell inequality* (see (51))

$$P(K_S, \bar{K}^0) \leq P(K_S, K_1^0) + P(K_1^0, \bar{K}^0). \tag{62}$$

Inequality (62) first considered by Uchiyama [65] can be converted into the inequality $\text{Re}\{\varepsilon\} \leq |\varepsilon|^2$ for the CP violation parameter ε, which is obviously violated by the experimental value of ε, having an absolute value of order 10^{-3} and a phase of about 45° [66].

We, however, want to stay as general and loophole-free as possible and demonstrate the relation of Bell inequalities to CP violation in the following way [67].

The Bell inequality (62) is rather formal because it involves the unphysical CP-even state $|K_1^0\rangle$, but it implies an inequality on a *physical CP violation* parameter which is experimentally testable. For the derivation, recall the H and CP eigenstates, (5) and (3), then we have the following transition amplitudes

$$\langle \bar{K}^0|K_S\rangle = -\frac{q}{N}, \quad \langle \bar{K}^0|K_1^0\rangle = -\frac{1}{\sqrt{2}}, \quad \langle K_S|K_1^0\rangle = \frac{1}{\sqrt{2}N}(p^* + q^*), \tag{63}$$

which we use to calculate the probabilities in BI (62). Optimizing the inequality we find, independent of any phase conventions of the kaon states,

$$|p| \leq |q|. \tag{64}$$

Proposition: p, q – kaon transition coefficients

- *Inequality $|p| \leq |q|$ is experimentally testable!*

Semileptonic Decays: Let us consider the semileptonic decays of the K mesons. The strange quark s decays weakly as constituent of $\bar{K}^0$:

Due to their quark content the kaon $K^0(\bar{s}d)$ and the anti-kaon $\bar{K}^0(s\bar{d})$ have the following definite decays:

decay of strange particles			quark level		
$K^0(d\bar{s}) \longrightarrow \pi^-(d\bar{u})\ \ l^+\ \ \nu_l$			$\bar{s} \longrightarrow \bar{u}\ \ l^+\ \ \nu_l$		
Q 0 −1			$\frac{1}{3}$ $-\frac{2}{3}$		
S 1 0			1 0		
$\bar{K}^0(\bar{d}s) \longrightarrow \pi^+(\bar{d}u)\ \ l^-\ \ \bar{\nu}_l$			$s \longrightarrow u\ \ l^-\ \ \bar{\nu}_l$		
Q 0 +1			$-\frac{1}{3}$ $\frac{2}{3}$		
S −1 0			−1 0		

$$\Delta S = \Delta Q \quad \text{rule}$$

In particular, we study the leptonic charge asymmetry

$$\delta = \frac{\Gamma(K_L \to \pi^- l^+ \nu_l) - \Gamma(K_L \to \pi^+ l^- \bar{\nu}_l)}{\Gamma(K_L \to \pi^- l^+ \nu_l) + \Gamma(K_L \to \pi^+ l^- \bar{\nu}_l)} \quad \text{with} \quad l = \mu, e \;, \quad (65)$$

where l represents either a muon or an electron. The $\Delta S = \Delta Q$ rule for the decays of the strange particles implies that – due to their quark content – the kaon $K^0(\bar{s}d)$ and the anti-kaon $\bar{K}^0(s\bar{d})$ have definite decays (see above table). Thus, l^+ and l^- tag K^0 and $\bar{K}^0$, respectively, in the K_L state, and the leptonic asymmetry (65) is expressed by the probabilities $|p|^2$ and $|q|^2$ of finding a K^0 and a $\bar{K}^0$, respectively, in the K_L state

$$\delta = \frac{|p|^2 - |q|^2}{|p|^2 + |q|^2} \;. \quad (66)$$

Then inequality (64) turns into the bound

$$\delta \leq 0 \quad (67)$$

for the leptonic charge asymmetry which measures CP violation.

If CP were conserved, we would have $\delta = 0$. Experimentally, however, the asymmetry is nonvanishing[1], namely

$$\delta = (3.27 \pm 0.12) \cdot 10^{-3} \;, \quad (68)$$

and is thus a clear sign of CP violation.

---

[1]It is the weighted average over electron and muon events, see [66].

The bound (67) dictated by BI (62) is in contradiction to the experimental value (68) which is definitely positive. *In this sense CP violation is related to the violation of a Bell inequality!*

On the other hand, we can replace $\bar{K}^0$ by K^0 in the BI (62) and along the same lines as discussed before we obtain the inequality

$$|p| \geq |q| \qquad \text{or} \qquad \delta \geq 0 . \tag{69}$$

Altogether inequalities (64), (67) and (69) imply the strict equality

$$|p| = |q| \qquad \text{or} \qquad \delta = 0 , \tag{70}$$

which is in contradiction to experiment.

Conclusion: The premises of LRT are *only* compatible with strict CP conservation in $K^0 \bar{K}^0$ mixing. Conversely, CP violation in $K^0 \bar{K}^0$ mixing, no matter which sign the experimental asymmetry (65) actually has, always leads to a *violation* of a BI, either of inequality (64), (67) or of (69). In this way, $\delta \neq 0$ is a manifestation of the entanglement of the considered state.

We also want to remark that in case of Bell inequality (62), since it is considered at $t = 0$, it is rather *contextuality* than nonlocality which is tested. *Noncontextuality* means that the value of an observable does not depend on the experimental context; the measurement of the observable must yield the value independent of other simultaneous measurements. The question is whether the properties of individual parts of a quantum system do have definite or predetermined values before the act of measurement – a main hypothesis in hidden variable theories. The *no-go theorem of Bell–Kochen–Specker* [68] states:

- *Noncontextual theories are incompatible with QM!*

The contextual quantum feature is verified in our case.

7 Decoherence in Entangled $K^0 \bar{K}^0$ System

Again, we consider the creation of an entangled kaon state at the Φ resonance; the state propagates to the left and right until the kaons are measured.

measure
quasi-spin: $|k_1\rangle_l$ on left side $\longleftrightarrow$ $|k_2\rangle_r$ on right side

How can we describe and measure possible decoherence in the entangled state? Decoherence provides us some information on the quality of the entangled state.

In the following we consider possible decoherence effects arising from some interaction of the quantum system with its "environment". Sources for "standard" decoherence effects are the strong interaction scatterings of kaons with nucleons, the weak interaction decays and the noise of the experimental setup. "Nonstandard" decoherence effects result from a fundamental modification of QM and can be traced back to the influence of quantum gravity [69, 70, 71] – quantum fluctuations in the space-time structure on the Planck mass scale – or to dynamical state-reduction theories [72, 73, 74, 75], and arise on a different energy scale. We will present in the following a specific model of decoherence and quantify the strength of such possible effects with the help of data of existing experiments.

7.1 Density Matrix

In decoherence theory there will occur a statistical mixture of quantum states, which can be elegantly described by a *density matrix*. It is of great importance for quantum statistics, therefore we briefly recall its conception and basic properties.

Usually a quantum system is described by a state vector $|\psi\rangle$ which is determined by the Schrödinger equation

$$i\frac{\partial}{\partial t}|\psi\rangle = H|\psi\rangle, \qquad \hbar = 1. \tag{71}$$

The expectation value of an observable A in the state $|\psi\rangle$ is calculated by

$$\langle A\rangle = \langle\psi|A|\psi\rangle. \tag{72}$$

Then it is rather suggestive to define a *density matrix for pure states* as

$$\rho = |\psi\rangle\langle\psi|, \tag{73}$$

with properties

$$\rho^2 = \rho, \qquad \rho^\dagger = \rho, \qquad \mathrm{tr}\rho = 1. \tag{74}$$

The extension to a statistical mixture of states with probabilities p_i – the *density matrix for mixed states* – is straight forward

$$\rho = \sum_i p_i|\psi_i\rangle\langle\psi_i| \qquad \text{with} \quad p_i \geq 0 \quad \sum_i p_i = 1. \tag{75}$$

The *mixed state density matrix* has the properties

$$\rho^2 \neq \rho, \qquad \rho^\dagger = \rho, \qquad \mathrm{tr}\rho = 1, \qquad \mathrm{tr}\rho^2 < 1. \tag{76}$$

Then the expectation value of observable A in a state ρ is defined by

$$\langle A \rangle = \operatorname{tr} \rho A \ . \tag{77}$$

Due to the Schrödinger equation (71) the time evolution of the density matrix is determined by an equation, called the *von Neumann equation*

$$i \frac{\partial}{\partial t} \rho = [H, \rho] \ . \tag{78}$$

Example: Density matrix for spin-$\frac{1}{2}$ state

$$\rho = \frac{1}{2}(\mathbb{1} + \boldsymbol{\rho} \cdot \boldsymbol{\sigma}) \quad \text{with} \quad \boldsymbol{\rho} = \operatorname{tr} \rho \boldsymbol{\sigma} = \langle \boldsymbol{\sigma} \rangle \quad \text{Bloch vector ;}$$

$$\text{if} \quad |\boldsymbol{\rho}| = 1 \quad \text{pure state ,}$$

$$\text{if} \quad |\boldsymbol{\rho}| < 1 \quad \text{mixed state .} \tag{79}$$

Explicitly, the density matrix for a pure state with spin $\boldsymbol{\sigma}$ along $\boldsymbol{\alpha}$ denotes

$$\rho = |\Uparrow \boldsymbol{\alpha}\rangle\langle\Uparrow \boldsymbol{\alpha}| = \begin{pmatrix} \cos^2 \frac{\alpha}{2} & \frac{1}{2}\sin\alpha\, e^{-i\phi} \\ \frac{1}{2}\sin\alpha\, e^{i\phi} & \sin^2 \frac{\alpha}{2} \end{pmatrix} \tag{80}$$

and for a mixed state with a 50 : 50 mixture of spin up and down we have

$$\rho_{\text{mixed}} = \frac{1}{2}\big(|\uparrow\rangle\langle\uparrow| + |\downarrow\rangle\langle\downarrow|\big) = \frac{1}{2}\mathbb{1} \ . \tag{81}$$

Proposition: for a density matrix of mixed states

- *There are different mixtures of states leading to the same ρ_{mixed}!*

Example:

$$\rho_{\text{mixed}} = \frac{1}{2}\,\uparrow\downarrow = \frac{1}{3}\,\measuredangle = \frac{1}{2}\,\mathbb{1}$$

Here, the up-down arrows denote the mixed state (81), where the weight is $\frac{1}{2}$, and the three star-like arrows represent a mixture of three states (80) with the angles $\alpha = 0°, \pm120°$ ($\phi = 0°$) and the weight $\frac{1}{3}$.

Physics:
- *The physics depends only on the density matrix ρ!*

$\Longrightarrow$ Several types of mixtures of the same ρ_{mixed} are not distinguishable. They are different expressions of incomplete information about system.

$\Longrightarrow$ The entropy of a quantum system measures the degree of uncertainty, i.e., the lack of knowledge, of the quantum state of a system.

7.2 Model

We discuss the model of decoherence in a 2-dimensional Hilbert space $\mathcal{H} = \mathbf{C}^2$ and consider the usual non-Hermitian "effective mass" Hamiltonian H which describes the decay properties and the strangeness oscillations of the kaons. The mass eigenstates, the short-lived $|K_S\rangle$ and long-lived $|K_L\rangle$ states, are determined by

$$H\,|K_{S,L}\rangle \;=\; \lambda_{S,L}\,|K_{S,L}\rangle \qquad \text{with} \quad \lambda_{S,L} \;=\; m_{S,L} - \frac{i}{2}\Gamma_{S,L}\,, \qquad (82)$$

with $m_{S,L}$ and $\Gamma_{S,L}$ being the corresponding masses and decay widths. For our purpose CP invariance[2] is assumed, i.e. the CP eigenstates $|K_1^0\rangle, |K_2^0\rangle$ are equal to the mass eigenstates

$$|K_1^0\rangle \equiv |K_S\rangle, \quad |K_2^0\rangle \equiv |K_L\rangle, \qquad \text{and} \quad \langle K_S|K_L\rangle = 0\,. \qquad (83)$$

As a starting point for our model of decoherence we consider the Liouville – von Neumann equation with the Hamiltonian (82) and allow for decoherence by adding a so-called *dissipator* $D[\rho]$, so that the time evolution of the density matrix ρ is governed by the following *master equation*

$$\frac{d\rho}{dt} \;=\; -iH\rho + i\rho H^\dagger - D[\rho]\,. \qquad (84)$$

For the dissipative term $D[\rho]$ we write the simple ansatz (see [42, 48])

$$D[\rho] \;=\; \lambda\left(P_S\rho P_L + P_L\rho P_S\right) \;=\; \frac{\lambda}{2}\sum_{j=S,L}\left[P_j, [P_j, \rho]\right]\,, \qquad (85)$$

where $P_j = |K_j\rangle\langle K_j|$ $(j = S, L)$ denote the projectors to the eigenstates of the Hamiltonian and λ is called *decoherence parameter*; $\lambda \geq 0$.

Remark: Note that we focus here on the undecayed kaon system which is described by the non-Hermitian "effective mass" Hamiltonian (6), (82). In this case the master equation (84) is not trace conserving. But it, clearly, becomes trace conserving again when we include the decay product states as well. So the total system is described by an enlarged Hilbert space being the direct sum of the kaon- and decay product space. Since our interest is the decoherence study of the undecayed K-meson system we confine ourselves to master equation (84) neglecting the part of the decay products.

Features: With ansatz (85) our model has the following nice features.

---

[2]Note that corrections due to CP violations are of order 10^{-3}, however, we compare this model of decoherence with the data of the CPLEAR experiment [35] which are not sensitive to CP violating effects.

(i)   It describes a completely positive map; when identifying $A_j = \sqrt{\lambda} P_j$, $j = S, L$, it is a special case of Lindblad's general structure [76]

$$D[\rho] = \frac{1}{2} \sum_j (A_j^\dagger A_j \, \rho + \rho A_j^\dagger A_j - 2 A_j \rho A_j^\dagger) \, . \qquad (86)$$

Equivalently, it is a special form of the Gorini–Kossakowski–Sudarshan expression [77] (see [78]).

(ii) It conserves energy in case of a Hermitian Hamiltonian since $[P_j, H] = 0$ (see [79]).

(iii) The von Neumann entropy $S(\rho) = -\,\mathrm{Tr}(\rho \ln \rho)$ is not decreasing as a function of time since $P_j^\dagger = P_j$, thus $A_j^\dagger = A_j$ in our case (see [80]).

With choice (85) the time evolution (84) decouples for the components of the matrix ρ, which are defined by

$$\rho(t) = \sum_{i,j=S,L} \rho_{ij}(t) \, |K_i\rangle\langle K_j| \, , \qquad (87)$$

and we obtain

$$\begin{aligned}
\rho_{SS}(t) &= \rho_{SS}(0) \cdot e^{-\Gamma_S t} \, , \\
\rho_{LL}(t) &= \rho_{LL}(0) \cdot e^{-\Gamma_L t} \, , \\
\rho_{LS}(t) &= \rho_{LS}(0) \cdot e^{-i\Delta m t - \Gamma t - \lambda t} \, ,
\end{aligned} \qquad (88)$$

where $\Delta m = m_L - m_S$ and $\Gamma = \frac{1}{2}(\Gamma_S + \Gamma_L)$.

7.3 Entangled Kaons

Let us study now entangled neutral kaons. We use the abbreviations

$$|e_1\rangle = |K_S\rangle_l \otimes |K_L\rangle_r \qquad \text{and} \qquad |e_2\rangle = |K_L\rangle_l \otimes |K_S\rangle_r \, , \qquad (89)$$

and regard – as usual – the total Hamiltonian as a tensor product of the 1-particle Hilbert spaces: $H = H_l \otimes 1_r + 1_l \otimes H_r$, where l denotes the left-moving and r the right-moving particle. The initial Bell singlet state

$$|\psi^-\rangle = \frac{1}{\sqrt{2}} \Big\{ |e_1\rangle - |e_2\rangle \Big\} \qquad (90)$$

is equivalently described by the density matrix

$$\rho(0) = |\psi^-\rangle\langle\psi^-| = \frac{1}{2} \Big\{ |e_1\rangle\langle e_1| + |e_2\rangle\langle e_2| - |e_1\rangle\langle e_2| - |e_2\rangle\langle e_1| \Big\} \, . \qquad (91)$$

Then the time evolution given by (84) with our ansatz (85), where now the operators $P_j = |e_j\rangle\langle e_j|$ $(j = 1, 2)$ project to the eigenstates of the 2-particle Hamiltonian H, also decouples

$$\rho_{11}(t) = \rho_{11}(0) \, e^{-2\Gamma t} \, ,$$
$$\rho_{22}(t) = \rho_{22}(0) \, e^{-2\Gamma t} \, ,$$
$$\rho_{12}(t) = \rho_{12}(0) \, e^{-2\Gamma t - \lambda t} \, . \tag{92}$$

Consequently, we obtain for the time-dependent density matrix

$$\rho(t) \;=\; \frac{1}{2} e^{-2\Gamma t} \Big\{ |e_1\rangle\langle e_1| + |e_2\rangle\langle e_2| - e^{-\lambda t} \big(|e_1\rangle\langle e_2| + |e_2\rangle\langle e_1| \big) \Big\} \, . \tag{93}$$

The decoherence arises through the factor $e^{-\lambda t}$ which only effects the off-diagonal elements. It means that for $t > 0$ and $\lambda \neq 0$ the density matrix $\rho(t)$ does not correspond to a pure state anymore.

Finally, in order to arrive at a proper density matrix for the kaon system, conditioned on having not decayed up to time t, we have to divide $\rho(t)$ (93) by the trace $\mathrm{Tr}\rho(t)$, see Sect. 9.1.

7.4 Measurement

In our model the parameter λ quantifies the strength of possible decoherence of the whole entangled state. We want to determine its permissible range of values by experiment.

Concerning the measurement, we have the following point of view. The 2-particle density matrix follows the time evolution given by (84) with the Lindblad generators $A_j = \sqrt{\lambda} \, |e_j\rangle\langle e_j|$ and undergoes thereby some decoherence. We measure the strangeness content S of the right-moving particle at time t_r and of the left-moving particle at time t_l. For sake of definiteness we choose $t_r \leq t_l$. For times $t_r \leq t \leq t_l$ we have a 1-particle state which evolves exactly according to QM, i.e. no further decoherence is picked up.

In theory we describe the measurement of the strangeness content, i.e. the right-moving particle being a K^0 or a $\bar{K}^0$ at time t_r, by the following projection onto ρ

$$\mathrm{Tr}_r \big\{ \mathbf{1}_l \otimes |S^{'}\rangle\langle S^{'}|_r \; \rho(t_r) \big\} \;\equiv\; \rho_l(t = t_r; t_r) \, , \tag{94}$$

where strangeness $S^{'} = +, -$ and $|+\rangle = |K^0\rangle$, $|-\rangle = |\bar{K}^0\rangle$. Consequently, $\rho_l(t; t_r)$ for times $t \geq t_r$ is the 1–particle density matrix for the left-moving particle and evolves as a 1–particle state according to pure QM. At $t = t_l$ the strangeness content ($S = +, -$) of the second particle is measured and we finally calculate the probability

$$P_\lambda(S, t_l; S^{'}, t_r) \;=\; \mathrm{Tr}_l \big\{ |S\rangle\langle S|_l \; \rho_l(t_l; t_r) \big\} \, . \tag{95}$$

Explicitly, we find the following results for the like-and unlike-strangeness probabilities

$$P_\lambda(K^0, t_l; K^0, t_r) = P_\lambda(\bar{K}^0, t_l; \bar{K}^0, t_r)$$
$$= \frac{1}{8}\left\{ e^{-\Gamma_S t_l - \Gamma_L t_r} + e^{-\Gamma_L t_l - \Gamma_S t_r} - e^{-\lambda t_r}\, 2\cos(\Delta m \Delta t) \cdot e^{-\Gamma(t_l + t_r)} \right\},$$

$$P_\lambda(K^0, t_l; \bar{K}^0, t_r) = P_\lambda(\bar{K}^0, t_l; K^0, t_r)$$
$$= \frac{1}{8}\left\{ e^{-\Gamma_S t_l - \Gamma_L t_r} + e^{-\Gamma_L t_l - \Gamma_S t_r} + e^{-\lambda t_r}\, 2\cos(\Delta m \Delta t) \cdot e^{-\Gamma(t_l + t_r)} \right\} \quad (96)$$

with $\Delta t = t_l - t_r$.

Note that at equal times $t_l = t_r = t$ the like-strangeness probabilities

$$P_\lambda(K^0, t; K^0, t) = P_\lambda(\bar{K}^0, t; \bar{K}^0, t) = \frac{1}{4}\, e^{-2\Gamma t}\, (1 - e^{-\lambda t}) \quad (97)$$

do not vanish, in contrast to the pure quantum mechanical EPR-correlations.

The interesting quantity is the *asymmetry of probabilities*; it is directly sensitive to the interference term and can be measured experimentally. For pure QM we have

$$A^{QM}(\Delta t)$$
$$= \frac{P(K^0, t_l; \bar{K}^0, t_r) + P(\bar{K}^0, t_l; K^0, t_r) - P(K^0, t_l; K^0, t_r) - P(\bar{K}^0, t_l; \bar{K}^0, t_r)}{P(K^0, t_l; \bar{K}^0, t_r) + P(\bar{K}^0, t_l; K^0, t_r) + P(K^0, t_l; K^0, t_r) + P(\bar{K}^0, t_l; \bar{K}^0, t_r)}$$
$$= \frac{\cos(\Delta m \Delta t)}{\cosh(\frac{1}{2}\Delta\Gamma \Delta t)}, \quad (98)$$

with $\Delta\Gamma = \Gamma_L - \Gamma_S$, and for our decoherence model we find, by inserting the probabilities (96),

$$A^\lambda(t_l, t_r) = \frac{\cos(\Delta m \Delta t)}{\cosh(\frac{1}{2}\Delta\Gamma \Delta t)}\, e^{-\lambda\,\min\{t_l, t_r\}}$$
$$= A^{QM}(\Delta t)\, e^{-\lambda\,\min\{t_l, t_r\}}. \quad (99)$$

Thus the decoherence effect, simply given by the factor $e^{-\lambda\,\min\{t_l, t_r\}}$, depends only – according to our philosophy – on the time of the first measured kaon, in our case: $\min\{t_l, t_r\} = t_r$.

7.5 Experiment

Now we compare our model with the results of the CPLEAR experiment [35] at CERN where $K^0 \bar{K}^0$ pairs are produced in the $p\bar{p}$ collider: $p\bar{p} \longrightarrow K^0 \bar{K}^0$. These pairs are predominantly in an antisymmetric state with quantum numbers $J^{PC} = 1^-$ and the strangeness of the kaons is detected via strong interactions in surrounding absorbers (made of copper and carbon).

Examples:

$$S = +1: \qquad K^0(d\bar{s}) + N \longrightarrow K^+(u\bar{s}) + X ,$$
$$S = -1: \qquad \bar{K}^0(\bar{d}s) + N \longrightarrow K^-(\bar{u}s) + X ,$$
$$S = -1: \qquad \bar{K}^0(\bar{d}s) + N \longrightarrow \Lambda(uds) + X \quad \text{and} \quad \Lambda \longrightarrow p\pi^- ;$$

like – strangeness events: (K^-, Λ) ,
unlike – strangeness events: (K^+, K^-), (K^+, Λ) .

The experimental set-up has two configurations (see Fig. 1). In configuration $C(0)$ both kaons propagate 2 cm, they have nearly equal proper times $(t_r \approx t_l)$ when they are measured by the absorbers. This fulfills the condition for an EPR-type experiment. In configuration $C(5)$ one kaon propagates 2 cm and the other kaon 7 cm, thus the flight–path difference is 5 cm on average, corresponding to a proper time difference $|t_r - t_l| \approx 1.2\tau_S$.

Fitting the decoherence parameter λ by comparing the asymmetry (99) with the experimental data [35] (see Fig. 2) we find, when averaging over both configurations, the following bounds on λ

$$\bar{\lambda} = (1.84^{+2.50}_{-2.17}) \cdot 10^{-12} \text{ MeV} \quad \text{and} \quad \bar{\Lambda} = \frac{\bar{\lambda}}{\Gamma_S} = 0.25^{+0.34}_{-0.32} . \qquad (100)$$

The results (100) are certainly compatible with QM ($\lambda = 0$), nevertheless, the experimental data allow an upper bound $\bar{\lambda}_{\text{up}} = 4.34 \cdot 10^{-12}$ MeV for possible decoherence in the entangled $K^0\bar{K}^0$ system.

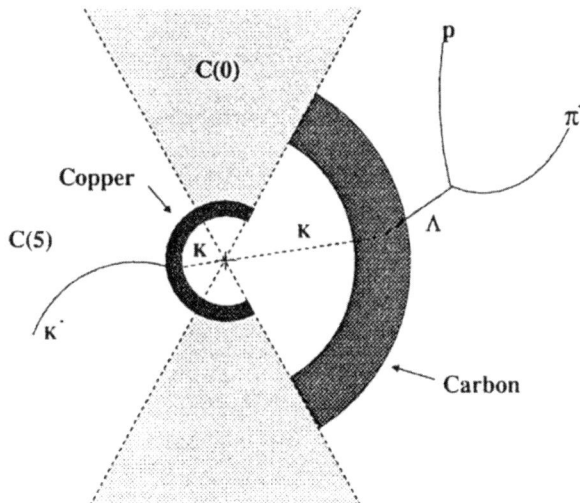

Fig. 1. Example of CPLEAR event: like-strangeness (K^-, Λ)

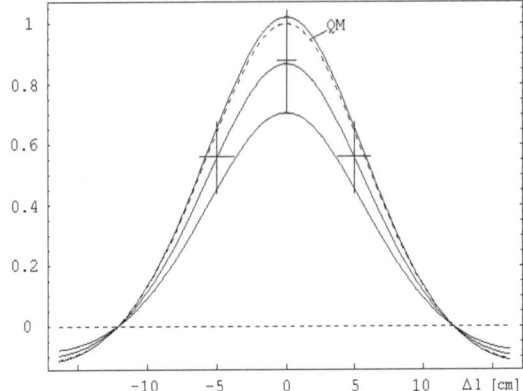

Fig. 2. The asymmetry (99) as a function of the flight-path difference of the kaons. The *dashed curve* corresponds to QM, the *solid curves* represent the best fit (100) to the CPLEAR data [35], given by the crosses. The *horizontal line* indicates the Furry–Schrödinger's hypothesis [1, 4], as explained in Sect. 8

$B^0\bar{B}^0$ **System:** An analogous investigation of "decoherence of entangled beauty" has been carried out in [42]. However, there time integrated events from the semileptonic decays of the entangled $B^0\bar{B}^0$ pairs are analyzed, which were produced at the colliders at DESY and Cornell. The analysis provides the bounds $\lambda_B = (-47 \pm 76) \cdot 10^{-12}$ MeV, which are an order of magnitude less restrictive than the bounds (100) of the $K^0\bar{K}^0$ system.

The possibility to measure the asymmetry at different times is offered now also in the B-meson system. Entangled $B^0\bar{B}^0$ pairs are created with high density at the asymmetric B-factories and identified by the detectors BELLE at KEK-B (see e.g. [81, 82]) and BABAR at PEP-II (see e.g. [83, 84]) with a high resolution at different distances or times.

8 Connection to Phenomenological Model

There exists a simple procedure [37, 38, 39] for introducing decoherence in a rather phenomenological way, more in the spirit of Furry [4] and Schrödinger [1] to describe the process of spontaneous factorization of the wavefunction. QM is modified in the sense that we multiply the interference term of the transition amplitude by the factor $(1 - \zeta)$. The quantity ζ is called *effective decoherence parameter*. Starting again from the Bell singlet state $|\psi^-\rangle$, which is given by the mass eigenstate representation (90), we have for the like-strangeness probability

$$P(K^0, t_l; K^0, t_r) = ||\langle K^0|_l \otimes \langle K^0|_r |\psi^-(t_l, t_r)\rangle||^2$$

$$\longrightarrow \quad P_\zeta(K^0, t_l; K^0, t_r) = \frac{1}{2} \left\{ e^{-\Gamma_S t_l - \Gamma_L t_r} |\langle K^0|K_S\rangle_l|^2 \, |\langle K^0|K_L\rangle_r|^2 \right.$$

$$+ e^{-\Gamma_L t_l - \Gamma_S t_r} |\langle K^0|K_L\rangle_l|^2 \, |\langle K^0|K_S\rangle_r|^2 \; - \; 2 \underbrace{(1-\zeta)}_{\text{modification}} e^{-\Gamma(t_l + t_r)}$$

$$\times \text{Re}\left\{ \langle K^0|K_S\rangle_l^* \langle K^0|K_L\rangle_r^* \langle K^0|K_L\rangle_l \langle K^0|K_S\rangle_r \, e^{-i\Delta m \Delta t} \right\}$$

$$= \frac{1}{8} \left\{ e^{-\Gamma_S t_l - \Gamma_L t_r} + e^{-\Gamma_L t_l - \Gamma_S t_r} \; - \; 2 \underbrace{(1-\zeta)}_{\text{modification}} e^{-\Gamma(t_l + t_r)} \cos(\Delta m \Delta t) \right\}, \tag{101}$$

and the unlike-strangeness probability just changes the sign of the interference term.

Features: The value $\zeta = 0$ corresponds to pure QM and $\zeta = 1$ to total decoherence or spontaneous factorization of the wave function, which is commonly known as Furry–Schrödinger hypothesis [1, 4]. The decoherence parameter ζ, introduced in this way "by hand", is quite effective [36, 37, 38, 39]; it interpolates continuously between the two limits: QM $\longleftrightarrow$ spontaneous factorization. It represents a measure for the amount of decoherence which results in a loss of entanglement of the total quantum state (we come back to this point in Sect. 9).

There exists a remarkable one–to–one correspondence between the model of decoherence (84), (85) and the phenomenological model, thus a relation between $\lambda \longleftrightarrow \zeta$ Refs. [42, 48]. We can see it quickly in the following way.

Calculating the asymmetry of strangeness events, as defined in (98), with the probabilities (101) we obtain

$$A^\zeta(t_l, t_r) = A^{QM}(\Delta t)\left(1 - \zeta(t_l, t_r)\right). \tag{102}$$

When we compare now the two approaches, i.e. (99) with (102), we find the formula

$$\zeta(t_l, t_r) = 1 - e^{-\lambda \min\{t_l, t_r\}}. \tag{103}$$

Of course, when fitting the experimental data with the two models, the λ values (100) are in agreement with the corresponding ζ values (averaged over both experimental setups) [36, 63],

$$\bar{\zeta} = 0.13^{+0.16}_{-0.15}. \tag{104}$$

The phenomenological model demonstrates that the $K^0 \bar{K}^0$ system is close to QM, i.e. $\zeta = 0$, *and* (at the same time) far away from total decoherence, i.e. $\zeta = 1$. It confirms nicely in a quantitative way the existence of entangled massive particles over macroscopic distances (9) cm.

We consider the decoherence parameter λ to be the fundamental constant, whereas the value of the effective decoherence parameter ζ depends on the time when a measurement is performed. In the time evolution of the state $|\psi^-\rangle$, (90), represented by the density matrix (93), we have the relation

$$\zeta(t) = 1 - e^{-\lambda t}, \tag{105}$$

which after measurement of the left- and right-moving particles at t_l and t_r turns into formula (103), if decoherence occurs as described in Sect. 7.4.

Our model of decoherence has a very specific time evolution (103). Measuring the strangeness content of the entangled kaons at definite times we have the possibility to distinguish experimentally, on the basis of time dependent event rates, between the prediction of our model (103) and the results of other models (which would differ from (103)). Indeed, it is of high interest to measure in future experiments the asymmetry of the strangeness events at various times, in order to confirm the very specific time dependence of the decoherence effect. In fact, such a possibility is now offered in the B-meson system; as we already mentioned entangled $B^0 \bar{B}^0$ pairs are created with high density at the asymmetric B-factories at KEK-B and at PEP-II.

9 Entanglement Loss – Decoherence

In the master equation (84) the *dissipator* $D[\rho]$ describes two phenomena occurring in an open quantum system, namely decoherence and dissipation (see, e.g., [85]). When the system S interacts with the environment E the initially product state evolves into entangled states of $S + E$ in the course of time [86, 87]. It leads to mixed states in S – which means decoherence – and to an energy exchange between S and E – which is called dissipation.

The decoherence destroys the occurrence of long-range quantum correlations by suppressing the off-diagonal elements of the density matrix in a given basis and leads to an information transfer from the system S to the environment E:

$$\boxed{S \rightleftarrows \quad E}$$

In general, both effects are present, however, decoherence acts on a much shorter time scale [86, 88, 89, 90] than dissipation and is the more important effect in quantum information processes.

Our model describes decoherence and not dissipation. The increase of decoherence of the initially totally entangled $K^0 \bar{K}^0$ system as time evolves means on the other hand a decrease of entanglement of the system. This loss of entanglement can be measured explicitly [48, 91].

In the field of quantum information the entanglement of a state is quantified by introducing certain *entanglement measures*. In this connection the entropy plays a fundamental role.

Entropy of a Quantum System:

- *The entropy measures the degree of uncertainty – the lack of knowledge – of a quantum state!*

In general, a quantum state can be in a pure or mixed state (see, our discussion in Sect. 7.1). Whereas the pure state supplies us with maximal information about the system, a mixed state does not.

Proposition: Thirring [92]

- *Mixed states provide only partial information about the system; the entropy measures how much of the maximal information is missing!*

For mixed states the entanglement measures cannot be defined uniquely. Common measures of entanglement are von Neumann's entropy function S, the entanglement of formation E and the concurrence C.

9.1 Von Neumann Entropy

We are only interested in the effect of decoherence, thus we properly normalize the state (93) in order to compensate for the decay property

$$\rho_N(t) = \frac{\rho(t)}{\text{Tr}\rho(t)} . \tag{106}$$

Von Neumann's entropy for the quantum state (106) is defined by

$$
\begin{aligned}
S(\rho_N(t)) &= -\,\text{Tr}\{\rho_N(t)\log_2\rho_N(t)\} \\
&= -\frac{1-e^{-\lambda t}}{2}\log_2\frac{1-e^{-\lambda t}}{2} - \frac{1+e^{-\lambda t}}{2}\log_2\frac{1+e^{-\lambda t}}{2} , \tag{107}
\end{aligned}
$$

where we have chosen the logarithm to base 2, the dimension of the Hilbert space (the qubit space), such that S varies between $0 \le S \le 1$.

Features:

(i) $S(\rho_N(t)) = 0$ for $t = 0$; the entropy is zero at time $t = 0$, there is no uncertainty in the system, the quantum state is pure and maximally entangled.

(ii) $S(\rho_N(t)) = 1$ for $t \to \infty$; the entropy increases for increasing t and approaches the value 1 at infinity. Hence the state becomes more and more mixed.

Reduced Density Matrices: Let us consider quite generally a composite quantum system A (Alice) and B (Bob). Then the reduced density matrix of subsystem A is given by tracing the density matrix of the joint state over all states of B.

In our case the subsystems are the propagating kaons on the left l and right r hand side thus we have as *reduced density matrices*

$$\rho_N^l(t) = \text{Tr}_r\{\rho_N(t)\} \quad \text{and} \quad \rho_N^r(t) = \text{Tr}_l\{\rho_N(t)\} . \tag{108}$$

The *uncertainty in the subsystem* l before the subsystem r is measured is given by the von Neumann entropy $S(\rho_N^l(t))$ of the corresponding reduced density matrix $\rho_N^l(t)$ (and alternatively we can replace $l \to r$). In our case we find

$$S\big(\rho_N^l(t)\big) = S\big(\rho_N^r(t)\big) = 1 \qquad \forall\, t \geq 0 . \tag{109}$$

The reduced entropies are independent of λ! That means the correlation stored in the composite system is, with increasing time, lost into the environment – what intuitively we had expected – and *not* into the subsystems, i.e. the individual kaons.

For pure quantum states von Neumann's entropy function (107) is a good measure for entanglement and, generally, A (Alice) and B (Bob) are most entangled when their reduced density matrices are maximally mixed.

For mixed states, however, von Neumann's entropy and the reduced entropies are no longer a good measure for entanglement so that we have to proceed in an other way to quantify entanglement (see Sect. 9.3).

9.2 Separability

In the following we want to show that the initially entangled Bell singlet state – although subjected to decoherence and thus to entanglement loss in the course of time – remains entangled. It is convenient to work with the "quasi-spin" description for the $K^0\bar{K}^0$ system (see Sect. 3). The projection operators of the mass eigenstates correspond to the spin projection operators "up" and "down"

$$P_S = |K_S\rangle\langle K_S| = \sigma_\uparrow = \frac{1}{2}(\mathbf{1}+\sigma_z) = \begin{pmatrix} 1 & 0 \\ 0 & 0 \end{pmatrix} ,$$

$$P_L = |K_L\rangle\langle K_L| = \sigma_\downarrow = \frac{1}{2}(\mathbf{1}-\sigma_z) = \begin{pmatrix} 0 & 0 \\ 0 & 1 \end{pmatrix} , \tag{110}$$

and the transition operators are the "spin-ladder" operators

$$P_{SL} = |K_S\rangle\langle K_L| = \sigma_+ = \frac{1}{2}(\sigma_x + i\,\sigma_y) = \begin{pmatrix} 0 & 1 \\ 0 & 0 \end{pmatrix} ,$$

$$P_{LS} = |K_L\rangle\langle K_S| = \sigma_- = \frac{1}{2}(\sigma_x - i\,\sigma_y) = \begin{pmatrix} 0 & 0 \\ 1 & 0 \end{pmatrix} . \tag{111}$$

Then density matrix (106), (93) is expressed by the Pauli spin matrices in the following way

$$\rho_N(t) = \frac{1}{4}\left\{ \mathbf{1} - \sigma_z \otimes \sigma_z - e^{-\lambda t}\left[\sigma_x \otimes \sigma_x + \sigma_y \otimes \sigma_y\right]\right\}, \qquad (112)$$

which at $t = 0$ coincides with the well-known expression for the pure spin singlet state $\rho_N(t = 0) = \frac{1}{4}\left(\mathbf{1} - \boldsymbol{\sigma} \otimes \boldsymbol{\sigma}\right)$; see, e.g., [93].

Operator (112) can be nicely written as 4×4 matrix

$$\rho_N(t) = \frac{1}{2}\begin{pmatrix} 0 & 0 & 0 & 0 \\ 0 & 1 & -e^{-\lambda t} & 0 \\ 0 & -e^{-\lambda t} & 1 & 0 \\ 0 & 0 & 0 & 0 \end{pmatrix}. \qquad (113)$$

For an other representation of the density matrix $\rho_N(t)$ we choose the so-called "Bell basis"

$$\rho^{\mp} = |\psi^{\mp}\rangle\langle\psi^{\mp}| \qquad \text{and} \qquad \omega^{\mp} = |\phi^{\mp}\rangle\langle\phi^{\mp}|, \qquad (114)$$

with $|\psi^-\rangle$ given by (90) and $|\psi^+\rangle$ by

$$|\psi^+\rangle = \frac{1}{\sqrt{2}}\left\{|e_1\rangle + |e_2\rangle\right\}. \qquad (115)$$

The states $|\phi^{\mp}\rangle = \frac{1}{\sqrt{2}}(|\uparrow\uparrow\rangle \mp |\downarrow\downarrow\rangle)$ (in spin notation) do not contribute here.

Entanglement – Separability

Recall that in general the density matrix ρ of a state is defined over the tensor product of Hilbert spaces $\mathcal{H} = \mathcal{H}_A \otimes \mathcal{H}_B$, named Alice and Bob.

A state ρ is then called *entangled* if it is *not separable*, i.e. $\rho \in S^c$ where S^c is the complement of the set of separable states S; and $S \cup S^c = \mathcal{H}$.

Separable States: The set of separable states is defined by

$$S = \left\{\rho = \sum_i p_i\, \rho_A^i \otimes \rho_B^i \;\Big|\; 0 \le p_i \le 1,\; \sum_i p_i = 1\right\}, \qquad (116)$$

where ρ_A^i and ρ_B^i are density matrices over the subspaces $\mathcal{H}_A$ and $\mathcal{H}_B$.

The important question is now to judge whether a quantum state is entangled or conversely separable or not. Several *separability criteria* give an answer to that.

Theorem 1. Positive partial transpose criterion, Peres–Horodecki [94, 95] Defining the partial transposition T_B by transposing only one of the subsystems, e.g. $T_B(\sigma^i)_{kl} = (\sigma^i)_{lk}$ in subsystem B, then a state ρ is separable iff its partial transposition with respect to any subsystem is positive:

$$(\mathbb{1}_A \otimes T_B)\,\rho \ge 0 \quad \text{and} \quad (T_A \otimes \mathbb{1}_B)\,\rho \ge 0 \quad \Longleftrightarrow \quad \rho \text{ separable}. \quad (117)$$

Theorem 2. Reduction criterion, Horodecki [96]
A state ρ is separable for:

$$\mathbb{1}_A \otimes \rho_B - \rho \geq 0 \quad and \quad \rho_A \otimes \mathbb{1}_B - \rho \geq 0 \quad \Longleftrightarrow \quad \rho \text{ separable}, \quad (118)$$

where ρ_A is Alice's reduced density matrix and ρ_B Bob's.

However, above Theorems (117), (118) are necessary and sufficient separability conditions – and so surprisingly simple – only for dimensions $2 \otimes 2$ and $2 \otimes 3$ [97]. A more general separability – entanglement criterion, valid in any dimensions, does exist; it is formulated by a so-called *generalized Bell inequality*, see [93].

Now let us return to the question of entanglement and separability of our kaon quantum state described by density matrix $\rho_N(t)$ (106), (93) as it evolves in time.

Proposition: Bertlmann–Durstberger–Hiesmayr [48]

- *The state represented by the density matrix $\rho_N(t)$ (106), (93) becomes mixed for $0 < t < \infty$ but remains entangled. Separability is achieved asymptotically $t \to \infty$ with the weight $e^{-\lambda t}$. Explicitly, $\rho_N(t)$ is the following mixture of the Bell states ρ^- and ρ^+ :*

$$\rho_N(t) = \frac{1}{2}\left(1 + e^{-\lambda t}\right)\rho^- + \frac{1}{2}\left(1 - e^{-\lambda t}\right)\rho^+ . \quad (119)$$

Proof:

(i) The mixedness of the state, with $t \to \infty$ totally mixed – separable, can be seen from the *mixed state criterion* (see Sect. 7.1)

$$\rho_N^2(t) = \frac{1}{4}\begin{pmatrix} 0 & 0 & 0 & 0 \\ 0 & 1 + e^{-2\lambda t} & -2e^{-\lambda t} & 0 \\ 0 & -2e^{-\lambda t} & 1 + e^{-2\lambda t} & 0 \\ 0 & 0 & 0 & 0 \end{pmatrix} \neq \rho_N(t) \text{ for } t > 0 . \quad (120)$$

(ii) Entanglement or lack of separability is determined by Theorems (117) and (118). The Peres–Horodecki partial transposition criterion (117)

$$(\mathbb{1}_l \otimes T_r)\,\rho_N(t) = \frac{1}{2}\begin{pmatrix} 0 & 0 & 0 & -e^{-\lambda t} \\ 0 & 1 & 0 & 0 \\ 0 & 0 & 1 & 0 \\ -e^{-\lambda t} & 0 & 0 & 0 \end{pmatrix} \ngeq 0 , \quad (121)$$

with eigenvalues $\left\{\frac{1}{2}, \frac{1}{2}, \frac{1}{2}e^{-\lambda t}, -\frac{1}{2}e^{-\lambda t}\right\}$, is not positive. Alternatively, the Horodecki reduction criterion (118), a matrix with same eigenvalues (121)

$$\mathbf{1}_l \otimes \rho_N^r(t) - \rho_N(t) \;=\; \frac{1}{2}\begin{pmatrix} 1 & 0 & 0 & 0 \\ 0 & 0 & e^{-\lambda t} & 0 \\ 0 & e^{-\lambda t} & 0 & 0 \\ 0 & 0 & 0 & 1 \end{pmatrix} \ngeq 0 \,, \qquad (122)$$

is not positive either. Therefore $\rho_N(t)$ remains entangled for $t < \infty$. $\square$

9.3 Entanglement of Formation and Concurrence

For pure states the entropy of the reduced density matrices is sufficient, for mixed states we need another measure, e.g., *entanglement of formation*.

Entanglement of Formation

Every density matrix ρ can be decomposed into an ensemble of pure states $\rho_i = |\psi_i\rangle\langle\psi_i|$ with the probability p_i, i.e. $\rho = \sum_i p_i \rho_i$. The entanglement of formation for a pure state is given by the entropy of either of the two subsystems. For a mixed state *entanglement of formation* [98] is defined as average entanglement of pure states of the decomposition, minimized over all decompositions of ρ

$$E(\rho) = \min \sum_i p_i \, S(\rho_i^l) \,. \qquad (123)$$

It quantifies the resources needed to create a given entangled state. Bennett et al. [98] found a remarkable simple formula for *entanglement of formation*

$$E(\rho) \geq \mathcal{E}\big(f(\rho)\big) \,, \qquad (124)$$

where the function $\mathcal{E}\big(f(\rho)\big)$ is defined by

$$\mathcal{E}(f) \;=\; H\left(\frac{1}{2} + \sqrt{f(1-f)}\right) \quad \text{for} \quad f \geq \frac{1}{2} \,, \qquad (125)$$

and $\mathcal{E}(f) = 0$ for $f < \frac{1}{2}$. The function H represents the familiar binary entropy function $H(x) = -x\log_2 x - (1-x)\log_2(1-x)$. The quantity $f(\rho)$ is called the *fully entangled fraction* of ρ

$$f(\rho) = \max \langle e|\rho|e\rangle \,, \qquad (126)$$

which is the maximum over all completely entangled states $|e\rangle$.

For general mixed states ρ the function $\mathcal{E}\big(f(\rho)\big)$ is only a lower bound to the entropy $E(\rho)$. For pure states and mixtures of Bell states – the case of our model – the bound is saturated, $E = \mathcal{E}$, and we have formula (125) for calculating the entanglement of formation.

Concurrence

Wootters and Hill [99, 100, 101] found that entanglement of formation for a general mixed state ρ of two qubits can be expressed by another quantity, the *concurrence* C

$$E(\rho) = \mathcal{E}(C(\rho)) = H\left(\frac{1}{2} + \frac{1}{2}\sqrt{1 - C^2}\right) \quad \text{with} \quad 0 \le C \le 1 . \quad (127)$$

Explicitly, the function $\mathcal{E}(C)$ looks like

$$\mathcal{E}(C) = -\frac{1 + \sqrt{1 - C^2}}{2} \log_2 \frac{1 + \sqrt{1 - C^2}}{2} - \frac{1 - \sqrt{1 - C^2}}{2} \log_2 \frac{1 - \sqrt{1 - C^2}}{2} \quad (128)$$

and is monotonically increasing from 0 to 1 as C runs from 0 to 1. Thus C itself is a kind of entanglement measure in its own right.

Defining the spin flipped state $\tilde{\rho}$ of ρ by

$$\tilde{\rho} = (\sigma_y \otimes \sigma_y) \rho^* (\sigma_y \otimes \sigma_y) , \quad (129)$$

where ρ^* is the complex conjugate and is taken in the standard basis, i.e. the basis $\{|\uparrow\uparrow\rangle, |\downarrow\downarrow\rangle, |\uparrow\downarrow\rangle, |\downarrow\uparrow\rangle\}$, the *concurrence* C is given by the formula

$$C(\rho) = \max\{0, \lambda_1 - \lambda_2 - \lambda_3 - \lambda_4\} . \quad (130)$$

The λ_i's are the square roots of the eigenvalues, in decreasing order, of the matrix $\rho\tilde{\rho}$.

Applications to Our Model

For the density matrix $\rho_N(t)$ (106) of our model, which is invariant under spin flip, i.e. $\tilde{\rho}_N = \rho_N$ and thus $\rho_N \tilde{\rho}_N = \rho_N^2$, we obtain for the concurrence

$$C(\rho_N(t)) = \max\{0, e^{-\lambda t}\} = e^{-\lambda t} , \quad (131)$$

and for the fully entangled fraction of $\rho_N(t)$

$$f(\rho_N(t)) = \frac{1}{2}(1 + e^{-\lambda t}) , \quad (132)$$

which is simply the largest eigenvalue of $\rho_N(t)$. Clearly, in our case the functions C and f are related by

$$C(\rho_N(t)) = 2 f(\rho_N(t)) - 1 . \quad (133)$$

Finally, we have for the entanglement of formation of the $K^0 \bar{K}^0$ system

$$E(\rho_N(t)) = -\frac{1 + \sqrt{1 - e^{-2\lambda t}}}{2} \log_2 \frac{1 + \sqrt{1 - e^{-2\lambda t}}}{2}$$
$$-\frac{1 - \sqrt{1 - e^{-2\lambda t}}}{2} \log_2 \frac{1 - \sqrt{1 - e^{-2\lambda t}}}{2} . \tag{134}$$

Using now our relation (105) between the decoherence parameters λ and ζ we find a striking connection between the entanglement measure, defined by the entropy of the state, and the decoherence of the quantum system, which describes the amount of factorization into product states (Furry–Schrödinger hypothesis [1, 4]).

Loss of Entanglement: Defining the *loss of entanglement* as the gap between an entanglement value and its maximum unity, we find

$$1 - C(\rho_N(t)) = \zeta(t) , \tag{135}$$

$$1 - E(\rho_N(t)) \doteq \frac{1}{\ln 2} \zeta(t) \doteq \frac{\lambda}{\ln 2} t , \tag{136}$$

where in (136) we have expanded expression (134) for small values of the parameters λ or ζ.

We get the following proposition.

Proposition: Bertlmann–Durstberger–Hiesmayr [48]

• *The entanglement loss equals the decoherence!*

Therefore we are able to determine experimentally the degree of entanglement of the $K^0 \bar{K}^0$ system, namely by considering the asymmetry (99) and fitting the parameter ζ or λ to the data.

Results

In Fig. 3 we have plotted the loss of entanglement $1 - E$, given by (134), as compared to the loss of information, the von Neumann entropy function S, (107), in dependence of the time t/τ_s of the propagating $K^0 \bar{K}^0$ system. The loss of entanglement of formation increases slower with time and visualizes the resources needed to create a given entangled state. At $t = 0$ the pure Bell state ρ^- is created and becomes mixed for $t > 0$ by the other Bell state ρ^+. In the total state the amount of entanglement decreases until separability is achieved (exponentially fast) for $t \to \infty$.

In case of the CPLEAR experiment, where one kaon propagates about 2 cm, corresponding to a propagation time $t_0/\tau_s \approx 0.55$, until it is measured by an absorber, the entanglement loss is about 18% for the mean value and maximal 38% for the upper bound of the decoherence parameter λ.

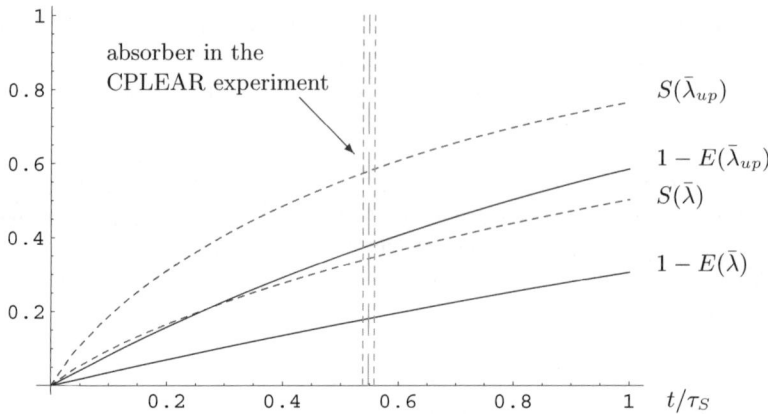

Fig. 3. The time dependence of the von Neumann entropy (*dashed lines*), (107), and the loss of entanglement of formation $1 - E$ (*solid lines*), given by (134), are plotted for the experimental mean value $\bar{\lambda} = 1.84 \cdot 10^{-12}$ MeV (*lower curve*) and the upper bound $\bar{\lambda}_{up} = 4.34 \cdot 10^{-12}$ MeV (*upper curve*), (100), of the decoherence parameter λ. The time t is scaled versus the lifetime τ_s of the short-lived kaon K_S: $t \to t/\tau_s$. The *vertical lines* represent the propagation time $t_0/\tau_s \approx 0.55$ of one kaon, including the experimental error bars, until it is measured by the absorber in the CPLEAR experiment

10 Outlook

Entanglement is the basic ingredient for quantum communication and computation, its effects will become important for future technologies. The fast developments in quantum information stimulated research in particle physics, which may have effects we cannot foresee now. We are sure that in future there will be a fruitful exchange between particle physics and quantum optics.

There are close analogies but also significant differences between entangled meson-antimeson systems and entangled photon or spin-$\frac{1}{2}$ systems. It turns out that quantum mechanical tests of meson-antimeson systems are more subtle than those of photon systems and one has to take into account the features of the mesons, which are characteristic for such massive quantum systems. It is the decay property or a symmetry violation property, like CP violation, or the regeneration property of the quantum state.

Bell Inequalities

Generally, Bell inequalities for meson-antimeson systems contain both the freedom of choice in time *and* in quasi-spin (see Sect. 6.2). As we concluded in Sect. 6.3 the time variation type of BI, however, cannot be used to exclude LRT for the familiar meson-antimeson systems due to the (unfortunate) interplay between flavor oscillation (e.g., strangeness, beauty, charm, ...) and decay time [47, 56, 62, 63].

In this connection we would like to mention that the work of [102], analyzing entangled $B^0 \bar{B}^0$ meson pairs produced at the KEKB asymmetric e^+e^- collider and collected at the BELLE detector, is hardly relevant for the test of QM versus LRT. The reasons are twofold: Firstly, "active" measurements are missing, therefore one can construct a local realistic model; secondly, the unitary time evolution of the unstable quantum state – the decay property of the meson – is ignored, which is part of its nature (for more detailed criticism, see [103]). Nevertheless, the work [102] represents a notable test of QM correlations exhibited by $B^0 \bar{B}^0$ entangled pairs and further investigations along these lines are recommended.

Considering, on the other hand, a BI at fixed time but for quasi-spin variation, a Wigner-like BI for the several types of the K-meson (see Sect. 6.4), provides an inequality for a symmetry violation parameter, the physical CP violation parameter (charge conjugation and parity) in our case [65, 67]. Experimentally it is tested by studying the leptonic charge asymmetry δ of the K_L type. It is remarkable that the premises of LRT are *only* compatible with strict CP conservation, i.e. with $\delta = 0$. In this way, $\delta \neq 0$ is a manifestation of the entanglement of the considered state. We have found the following proposition.

Proposition: Bertlmann–Grimus–Hiesmayr [67]

- CP *violation in* $K^0 \bar{K}^0$ *mixing leads to violation of a BI!*

We do believe that this connection between symmetry violation and BI violation is part of a more general quantum feature, therefore studies of other particle symmetries in this connection would be of high interest.

Bramon et al. [30, 31, 32] have established novel BI's for entangled $K^0 \bar{K}^0$ pairs by using the well-known regeneration mechanism of the kaons (see Sect. 2.4). After producing kaon pairs at the Φ resonance a thin regenerator (with kaon crossing time $t_{\text{cross}} \ll \tau_S$) is placed into one kaon beam near the Φ decay point. Then the entangled state contains a term proportional to $r |K_L\rangle_l \otimes |K_L\rangle_r$, where r denotes the *regeneration parameter*, a well-defined quantity being proportional to the difference of the $K^0 N$ and $\bar{K}^0 N$ amplitudes. Allowing the entangled state to propagate up to a time T, with $\tau_S \ll T \ll \tau_L$, in each beam of the two sides either the lifetime/decay states K_S versus K_L or the strangeness states K^0 versus $\bar{K}^0$ are measured. Considering *Clauser–Horne–type inequalities* [104] a set of inequalities for the parameter $R = -r \exp\left[-i\Delta m + \frac{1}{2}(\Gamma_S - \Gamma_L)\right] T$ can be derived. One of the inequalities is – under certain conditions (for the regeneration, etc...) – violated by QM. It provides an interesting experimental test at Φ-factories or $p\bar{p}$ machines in the future.

Decoherence

We have developed a general but quite simple and practicable procedure to estimate quantitatively the degree of possible decoherence of a quantum state due to some interaction of the state with its "environment" (see Sect. 7). This is of special interest in case of entangled states where a single parameter, the decoherence parameter λ or ζ [36, 42, 48], quantifies the strength of decoherence between the two subsystems or the amount of spontaneous factorization of the wavefunction. The asymmetry of like- and unlike-flavor events is directly sensitive to λ or ζ.

On the other hand, we have found a bridge to common entanglement measures, such as entanglement of formation and concurrence (see Sect. 9), and could derive the Proposition that *the entanglement loss equals the decoherence*! In this way we also measure the amount of entanglement of the state.

Using experimental data of the CPLEAR experiment provides us with upper bounds to the decoherence and the entanglement loss of the $K^0\bar{K}^0$ state produced in $p\bar{p}$ collisions, which is of macroscopic extent (9 cm).

As up to now only two data points are available, further experimental data are certainly highly desirable and would sharpen the bounds considerably. Such data could be collected at the Φ-factory DAΦNE in Frascati. In this connection, a recent analysis of entangled $K_S K_L$ systems by the KLOE collaboration [105] sets new bounds on the decoherence parameter ζ.

Indeed, it would be of great interest to measure in future experiments the asymmetry of like- and unlike-flavor events for several different times, in order to confirm the time dependence of the decoherence effect. In fact, such a possibility is now offered in the B-meson system. As already mentioned (Sect. 7.4) entangled $B^0\bar{B}^0$ pairs are created with high density at the asymmetric B-factories and identified by the detectors BELLE at KEK-B [81, 82] and BABAR at PEP-II [83, 84] with a high resolution at different distances or times. On the basis of the time dependent event rates we could test experimentally the predictions of our decoherence model.

Of course, these decoherence investigations could be performed with other entangled systems, like photonic systems, as well.

Complementarity

Meson systems in particle physics are also an interesting testing ground for basic principles of QM such as *quantitative wave-particle duality* [106, 107] or *quantum marking* and *quantum erasure* [108, 109].

Bohr's *complementarity principle* (or *duality principle*) can be formulated in a quantitative way [110, 111] by defining on one hand the *fringe visibility* – the sharpness of the interference pattern, the wave-like property – and on the other the *path predictability*, e.g., in a two-path interferometer [112]. This "which-way" knowledge on the path taken by the interfering system expresses the particle-like property of the system.

Considering now a produced K^0 beam, it oscillates between K^0 and $\bar{K}^0$ but it also represents a superposition of K_S and K_L which decay at totally different rates (recall Sect. 2). Then the strangeness oscillations can be viewed as fringe visibility and the K_S and K_L states with their different decay distances as the analogues of the two paths in the interferometer. The comparison of present data with the calculations of this kind of *kaon interferometry* satisfies nicely the statement of *quantitative duality* [106, 107]. Further experiments, e.g., at the Φ-factory DAΦNE are of interest in this connection.

Geometric Phase – Entanglement

An interesting question is: how can we influence the entanglement of a spin-$\frac{1}{2}$ system by creating a geometric phase? Geometric phases such as the Berry phase [113] play a considerable role in physics and arise in a quantum system when its time evolution is cyclic and adiabatic. There is increasing interest in combining both the geometric phase and the entanglement of a system [114, 115, 116, 117, 118].

In Ref. [118] we have studied the influence of the Berry phase on the entanglement of a spin-$\frac{1}{2}$ system by generating the Berry phase with an adiabatically rotating magnetic field in one of the paths of the particles, and we have eliminated the dynamical phase – being sensitive just to the geometric phase – by a "spin-echo" method. We have considered a *pure* entangled system where a phase – like our geometrical one – does *not* change the amount of entanglement and therefore *not* the extent of nonlocality of the system, which is determined by the maximal violation of a BI. In our case the Berry phase just affects the Bell angles in a specific way. On the other hand, keeping the measurement planes fixed the polar Bell angles vary and the maximum of the S-function of the CHSH inequality (recall Sect. 5) varies with respect to the Berry phase.

We have applied the investigation to neutron interferometry where one can achieve entanglement between different degrees of freedom – an internal and external one – the spin and the path of the neutron. In this case it is physically rather contextuality than nonlocality which is tested experimentally [112, 118, 119, 120].

There are still open problems in how does entanglement behave – the several entanglement measures – if geometric phases are introduced into *mixed* quantum states. How do they change entanglement? What happens to the Berry phase of the system if in addition we allow for interaction of the quantum system with an environment, thus allowing for decoherence? How can it be experimentally realized? Work along these lines is of increasing interest [121, 122, 123, 124, 125].

Generalized Bell Inequality – Entanglement Witness

In quantum communication, the two remote parties, Alice and Bob, want to carry out a certain task with minimal communication. Comparing now the amount of communication necessary by using classical *or* quantum bits, it turns out that sharing entangled quantum states – instead of classical or separable ones – leads to savings in the communication. That makes entangled states so important in quantum communication technology.

But how do we witness an entangled state? Via the familiar Bell inequalities described before (Sect. 5)? *Yes*, if the states are *pure* entangled states but *no* if they are *mixed* entangled. Werner [126] found out that a certain mixture of entangled and separable states, so-called *Werner states*, do satisfy the familiar BI. Thus the importance of the BI is that they serve as criterion for nonlocality – as departures from LRT – and they are valuable when considering phenomena like quantum teleportation or quantum cryptography. But as a criterion for separability they are rather poor.

There exist, however, some operators, so-called *entanglement witnesses*, satisfying an inequality, a *generalized Bell inequality* (GBI), for *all* separable states and violating this inequality for an entangled state [93, 127, 128, 129].

In Ref. [93] we have studied the class of entanglement witnesses; considering density matrices (states) and operators (entanglement witnesses) as elements of a Hilbert space it turns out that entanglement witnesses are tangent functionals to the set of separable states. Considering the Euclidean distance of the vectors in Hilbert space for entangled and separable states we have found the following Theorem.

Theorem: Bertlmann–Narnhofer–Thirring [93]

- *The Euclidean distance of an entangled state to the separable states is equal to the maximal violation of the GBI with the tangent functional as entanglement witness!*

Viewing the Euclidean distance as entanglement measure we have found a nice geometric picture for entanglement (and its value), for GBI (and its maximal violation) and for the tangent functional which characterizes (as a tangent) the set of separable states. All three conceptions are only different aspects of the same geometric picture. It is especially illustrative for the example of two spins, Alice and Bob. In this connection investigations for multi-particle entangled states, e.g., GHZ-states [130, 131], are certainly of high interest.

Acknowledgement

The author wants to take the opportunity to thank his collaborators and friends for all the joyful discussions about this field, special thanks go to

Markus Arndt, Katharina Durstberger, Gerhard Ecker, Stefan Filipp, Walter Grimus, Yuji Hasegawa, Beatrix Hiesmayr, Heide Narnhofer, Helmut Neufeld, Herbert Pietschmann, Helmut Rauch, Walter Thirring and Anton Zeilinger. The aid of EU project EURIDICE EEC-TMR program HPRN-CT-2002-00311 is acknowledged. Finally, I would like to thank the organizers of the Schladming School for creating such a stimulating scientific atmosphere.

References

1. E. Schrödinger: Naturwissenschaften **23**, 807 (1935); **23**, 823 (1935); **23**, 844 (1935)
2. A. Einstein, B. Podolski, N. Rosen: Phys. Rev. **47**, 777 (1935)
3. N. Bohr: Phys. Rev. **48**, 696 (1935)
4. W.H. Furry: Phys. Rev. **49**, 393 (1936); ibid. **49**, 476 (1936)
5. J.S. Bell: Physics **1**, 195 (1964)
6. R.A. Bertlmann: Found. Phys. **20**, 1191 (1990)
7. S.J. Freedman, J.F. Clauser: Phys. Rev. Lett. **28**, 938 (1972)
8. E.S. Fry, R.C. Thompson: Phys. Rev. Lett. **37**, 465 (1976)
9. A. Aspect, P. Grangier, G. Roger: Phys. Rev. Lett. **49**, 91 (1982); A. Aspect, J. Dalibard, G. Roger: Phys. Rev. Lett. **49**, 1804 (1982)
10. G. Weihs, T. Jennewein, C. Simon, H. Weinfurter, A. Zeilinger: Phys. Rev. Lett. **81**, 5039 (1998)
11. W. Tittel, G. Ribordy, N. Gisin: Physics World **11**, No. 3, p. 41 (1998); W. Tittel, J. Brendel, H. Zbinden, N. Gisin: Phys. Rev. Lett. **81**, 3563 (1998); H. Zbinden, J. Brendel, W. Tittel, N. Gisin: Phys. Rev. A **63**, 022111 (2001)
12. M. Aspelmeyer, H.R. Böhm, T. Gyatso, T. Jennewein, R. Kaltenbaek, M. Lindenthal, G. Molina-Terriza, A. Poppe, K. Resch, M. Taraba, R. Ursin, P. Walther, A. Zeilinger: Science **301**, 621 (2003)
13. J.S. Bell: *Speakable and Unspeakable in Quantum Mechanics* (Cambridge University Press 1987)
14. A.K. Ekert: Phys. Rev. Lett. **67**, 661 (1991)
15. D. Deutsch, A.K. Ekert: Physics World **11**, No. 3, p. 47 (1998)
16. R.J. Hughes: Contemp. Phys. **36**, 149 (1995)
17. C.H. Bennett, G. Brassard, C. Crépeau, R. Jozsa, A. Peres, W.K. Wootters: Phys. Rev. Lett. **70**, 1895 (1993)
18. J.-W. Pan, D. Bouwmeester, H. Weinfurter, A. Zeilinger: Nature **390**, 575 (1997)
19. A. Zeilinger: Physics World **11**, No. 3, 35 (1998)
20. D. Bouwmeester, A. Ekert, A. Zeilinger: *The physics of quantum information: quantum cryptography, quantum teleportation, quantum computations* (Springer, Berlin Heidelberg New York 2000)
21. R.A. Bertlmann, A. Zeilinger (eds.): *Quantum [Un]speakables, from Bell to Quantum Information* (Springer, Berlin Heidelberg New York 2002)
22. T.D. Lee, C.N. Yang: reported by T.D. Lee at Argonne National Laboratory, May 28, 1960 (unpublished)
23. D.R. Inglis: Rev. Mod. Phys. **33**, 1 (1961)
24. T.B. Day: Phys. Rev. **121**, 1204 (1961)

25. H.J. Lipkin: Phys. Rev. **176**, 1715 (1968)
26. P.H. Eberhard: in *The Second DaΦne Physics Handbook*, edited by L. Maiani, G. Pancheri, N. Paver (SIS–Pubblicazioni dei Laboratori di Frascati, Italy, 1995) Vol I, p. 99
27. P.H. Eberhard: Nucl. Phys. B **398**, 155 (1992)
28. I.I. Bigi: Nucl. Phys. B (Proc. Suppl.) **24A**, 24 (1991)
29. A. Di Domenico: Nucl. Phys. B **450**, 293 (1995)
30. A. Bramon, M Nowakowski: Phys. Rev. Lett. **83**, 1 (1999)
31. B. Ancochea, A. Bramon, M Nowakowski: Phys. Rev. D **60**, 094008 (1999)
32. A. Bramon, G. Garbarino: Phys. Rev. Lett. **88**, 040403 (2002)
33. M. Genovese, C. Novero, E. Predazzi: Phys. Lett. B **513**, 401 (2001)
34. J. Six: Phys. Lett. B **114**, 200 (1982)
35. A. Apostolakis et al.: CPLEAR Coll., Phys. Lett. B **422**, 339 (1998)
36. R.A. Bertlmann, W. Grimus, B.C. Hiesmayr: Phys. Rev. D **60**, 114032 (1999)
37. R.A. Bertlmann, W. Grimus: Phys. Lett. B **392**, 426 (1997)
38. G.V. Dass, K.V.L. Sarma: Eur. Phys. J. C **5**, 283 (1998)
39. R.A. Bertlmann, W. Grimus: Phys. Rev. D **58**, 034014 (1998)
40. A. Datta, D. Home: Phys. Lett. A **119**, 3 (1986)
41. A. Pompili, F. Selleri: Eur. Phys. J. C **14**, 469 (2000)
42. R.A. Bertlmann, W. Grimus: Phys. Rev. D **64**, 056004 (2001)
43. F. Selleri: Lett. Nuovo Cim. **36**, 521 (1983); P. Privitera, F. Selleri: Phys. Lett. B **296**, 261 (1992); F. Selleri: Phys. Rev. A **56**, 3493 (1997); R. Foadi, F. Selleri: Phys. Rev. A **61**, 012106 (1999); R. Foadi, F. Selleri: Phys. Lett. B **461**, 123 (1999)
44. A. Afriat, F. Selleri: *The Einstein, Podolsky, and Rosen Paradox in Atomic, Nuclear and Particle Physics*, (Plenum Press, New York London 1999)
45. J. Six: Phys. Lett. A **150**, 243 (1990)
46. A. Bramon, G. Garbarino: Phys. Rev. Lett. **89**, 160401 (2002)
47. R.A. Bertlmann, B. C. Hiesmayr: Phys. Rev. A **63**, 062112 (2001)
48. R.A. Bertlmann, K. Durstberger, B.C. Hiesmayr: Phys. Rev. A **68**, 012111 (2003)
49. M. Gell-Mann, A. Pais: Phys. Rev. **97**, 1387 (1955)
50. A. Pais, O. Piccioni: Phys. Rev. **100**, 1487 (1955)
51. J.S. Bell: Proceedings of the Royal Society A **231**, 479 (1955)
52. W. Pauli: *Niels Bohr and the development of physics*, (McGraw-Hill, New York 1955); Nuovo Cimento **6**, 204 (1957)
53. G. Lüders: Kgl. Danske Vidensk. Selsk. Mat.-Fys. Medd **28**, no. 5 (1954); Ann. Phys. **2**, 1 (1957)
54. T.D. Lee, C.S. Wu: Ann. Rev. Nucl. Sci. **16**, 511 (1966)
55. J.S. Bell, J. Steinberger: *Weak Interactions of Kaons*, Proceedings of the Oxford International Conference, 19-25 Sept., p. 147, (1965)
56. G.C. Ghirardi, R. Grassi, T. Weber: in Proc. of Workshop on *Physics and Detectors for DAΦNE, the Frascati Φ Factory*, April 9-12th, 1991, edited by G. Pancheri, p. 261
57. J.E. Clauser, M.A. Horne, A. Shimony, R.A. Holt: Phys. Rev. D **10**, 880 (1969)
58. J.S. Bell: Introduction to the Hidden–Variable Question. Article 4 in [13]
59. J.E. Clauser, A. Shimony: Rep. Prog. Phys. Vol. **41**, 1881 (1978)
60. E.P. Wigner: Am. J. Phys. **38**, 1005 (1970)
61. N. Gisin, A. Go: Am. J. Phys. **69** (3), 264 (2001)

62. G.C. Ghirardi, R. Grassi, R. Ragazzon: in *The DAΦNE Physics Handbook*, Vol. I, ed. by L. Maiani, G. Pancheri and N. Paver (Servizio Documentazione dei Laboratori Nazionale di Frascati, Italy 1992), p. 283

63. B.C. Hiesmayr: *The puzzling story of the $K^0 \bar{K}^0$-system or about quantum mechanical interference and Bell inequalities in particle physics*, Master Thesis, University of Vienna, (1999)

64. B.C. Hiesmayr: private communication

65. F. Uchiyama: Phys. Lett. A **231**, 295 (1997)

66. D.E. Groom et al.: *Review of Particle Physics*, Eur. Phys. J. C **3**, 1 (1998)

67. R.A. Bertlmann, W. Grimus, B.C. Hiesmayr: Phys. Lett. A **289**, 21 (2001)

68. J.S. Bell: Rev. Mod. Phys. **38**, 447 (1966); S. Kochen, E.P. Specker: J. Math. Mech. **17**, 59 (1967); N.D. Mermin: Rev. Mod. Phys. **65**, 803 (1993)

69. S. Hawking: Commun. Math. Phys. **87**, 395 (1982)

70. G. 't Hooft: Class. Quant. Grav. **16**, 3263 (1999)

71. G. 't Hooft: *Determinism and Dissipation in Quantum Gravity*, hep-th/0003005 (2000)

72. G.C. Ghirardi, A. Rimini, and T. Weber: Phys. Rev. D **34**, 470 (1986)

73. P. Pearle: Phys. Rev. A **39**, 2277 (1989)

74. N. Gisin, I.C. Percival: J. Phys. A: Math. Gen. **25**, 5677 (1992); *ibid.* **26**, 2245 (1993)

75. R. Penrose: Gen. Rel. Grav. **28**, 581 (1996); Phil. Trans. Roy. Soc. Lond. A **356**, 1927 (1998)

76. G. Lindblad: Comm. Math. Phys. **48**, 119 (1976)

77. V. Gorini, A. Kossakowski, E.C.G. Sudarshan: J. Math. Phys. **17**, 821 (1976)

78. R.A. Bertlmann, W. Grimus: Phys. Lett. A **300**, 107 (2002)

79. S.L. Adler: Phys. Rev. D **62**, 117901 (2000)

80. F. Benatti, H. Narnhofer: Lett. Math. Phys. **15**, 325 (1988)

81. BELLE-homepage: http://belle.kek.jp

82. G. Leder: talk given at the conference SUSY'02, Hamburg June 2002, http://wwwhephy.oeaw.ac.at/p3w/belle/leder_susy02.pdf (unpublished)

83. B. Aubert et al.: BABAR Collaboration, Phys. Rev. D. **66**, 032003 (2002)

84. BABAR-homepage: http://www.slac.stanford.edu/babar

85. H.-P. Breuer, F. Petruccione: *The theory of open quantum systems*, (Oxford University Press 2002)

86. E. Joos: *Decoherence through interaction with the environment*, in: *Decoherence and the apperance of a classical world in quantum theory*, D. Giulini et al. (eds.) (Springer Verlag, Heidelberg, 1996) p. 35

87. O. Kübler, H.D. Zeh: Ann. Phys. (N.Y.) **76**, 405 (1973)

88. E. Joos, H.D. Zeh: Z. Phys. B **59**, 223 (1985)

89. W.H. Zurek: Physics Today **44**, 36 (1991)

90. R. Alicki: Phys. Rev. A **65**, 034104 (2002)

91. B.C. Hiesmayr: *The puzzling story of the neutral kaon system or what we can learn of entanglement*, Ph.D. Thesis, University of Vienna, 2002

92. W. Thirring: *Lehrbuch der Mathematischen Physik 4, Quantenmechanik großer Systeme* (Springer Verlag, Wien 1980)

93. R.A. Bertlmann, H. Narnhofer, W. Thirring: Phys. Rev. A **66**, 032319 (2002)

94. A. Peres: Phys. Rev. Lett. **77**, 1413 (1996)

95. M. Horodecki, P. Horodecki, R. Horodecki: Phys. Lett. A **223**, 1 (1996)

96. M. Horodecki, P. Horodecki: Phys. Rev. A **59**, 4206 (1999)

97. M. Horodecki, P. Horodecki, R. Horodecki: *Mixed state entanglement and quantum communication*, in: *Quantum information*, G. Alber et al. (eds.), Springer Tracts in Modern Physics **173** (Springer Verlag Berlin 2001) p. 151
98. C.H. Bennett, D.P. DiVincenzo, J.A. Smolin, W.K. Wootters: Phys. Rev. A **54**, 3824 (1996)
99. S. Hill, W.K. Wootters: Phys. Rev. Lett. **78**, 5022 (1997)
100. W.K. Wootters: Phys. Rev. Lett. **80**, 2245 (1998)
101. W.K. Wootters: Quantum Information and Computation **1**, 27 (2001)
102. A. Go: J. Mod. Optics **51**, 991 (2004)
103. R.A. Bertlmann, A. Bramon, G. Garbarino, B.C. Hiesmayr: *Violation of a Bell inequality in particle physics experimentally verified?*, quant-ph/0409051 (2004)
104. J.F. Clauser, M.A. Horne: Phys. Rev. **10**, 526 (1974)
105. A. Di Domenico: KLOE Coll., *Kaon interferometry at KLOE: present and future*, hep-ex/0312032 (2003)
106. A. Bramon, G. Garbarino, B.C. Hiesmayr: Eur. J. Phys. C **32**, 377 (2004)
107. A. Bramon, G. Garbarino, B.C. Hiesmayr: Phys. Rev. A **69**, 022112 (2004)
108. A. Bramon, G. Garbarino, B.C. Hiesmayr: Phys. Rev. Lett. **92**, 020405 (2004)
109. A. Bramon, G. Garbarino, B.C. Hiesmayr: Phys. Rev. A **69**, 062111 (2004)
110. D. Greenberger, A. Yasin: Phys. Lett. A **128**, 391 (1988)
111. B.-G. Englert: Phys. Rev. Lett. **77**, 2154 (1996)
112. H. Rauch, S.A. Werner: *Neutron Interferometry* (Oxford Univ. Press, Oxford 2000)
113. M.V. Berry: Proc. R. Soc. Lond. A **392**, 45 (1984)
114. E. Sjöqvist: Phys. Rev. A **62**, 022109 (2000)
115. D.M. Tong, L.C. Kwek, C.H. Oh: J. Phys. A: Math. Gen. **36**, 1149 (2003)
116. P. Milman, R. Mosseri: *Topological phase for entangled two-qubit states*, quant-ph/0302202 (2003)
117. D.M. Tong, E. Sjöqvist, L.C. Kwek, C.H. Oh, M. Ericsson: Phys. Rev. A **68**, 022106 (2003)
118. R.A. Bertlmann, K. Durstberger, Y. Hasegawa, B.C. Hiesmayr: Phys. Rev. A **69**, 032112 (2004)
119. S. Basu, S. Bandyopadhyay, G. Kar, D. Home: Phys. Lett. A **279**, 281 (2001)
120. Y. Hasegawa, R. Loidl, G. Badurek, M. Baron, H. Rauch: Nature **425**, 45 (2003)
121. K.M. Fonseca Romero, A.C. Aguiar Pinto, M.T. Thomaz: Physica A **307**, 142 (2002)
122. I. Kamleitner, J.D. Cresser, B.C. Sanders: *Geometric phase for an adiabatically evolving open quantum system*, quant-ph/0406018 (2004)
123. A. Carollo, I. Fuentes-Guridi, M. França Santos, V. Vedral: *Geometric phase in open systems*, quant-ph/0301037 (2003)
124. P.J. Dodd, J.J. Halliwell: Phys. Rev. A **69**, 052105 (2004)
125. R.S. Whitney, Y. Gefen: *Berry phase in a non-isolated system*, cond-mat/0208141 v2 (2003)
126. R.F. Werner: Phys. Rev. A **40**, 4277 (1989)
127. M. Horodecki, P. Horodecki, R. Horodecki: Phys. Lett. A **223**, 1 (1996)
128. M.B. Terhal: Phys. Lett. A **271**, 319 (2000); M.B. Terhal: *Detecting quantum entanglement*, quant-ph/0101032 (2001)
129. D. Bruß: *Characterizing entanglement*, quant-ph/0110078 (2001)

130. D.M. Greenberger, M.A. Horne, A. Zeilinger: *Going beyond Bell's theorem*, in *Bell's Theorem, Quantum Theory and Conceptions of the Universe*, M. Kafatos (ed.), (Kluwer, Dordrecht 1989), pp. 73–76
131. D.M. Greenberger, M.A. Horne, A. Shimony, A. Zeilinger: Am. J. Phys. **58** (12), 1131 (1990)

Quantum Gates and Decoherence

S. Scheel[1], J.K. Pachos[2], E.A. Hinds[1], and P.L. Knight[1]

[1] Quantum Optics and Laser Science, Blackett Laboratory, Imperial College London, Prince Consort Road, London SW7 2BW, United Kingdom
s.scheel@imperial.ac.uk
[2] DAMTP, Centre for Mathematical Sciences, Wilberforce Road, Cambridge CB3 0WA, United Kingdom.

> "God forbid that we should give out a dream of our own imagination for a Pattern of the World."
> — Francis Bacon, Novum Organum

Abstract. In this article we are concerned with some possible physical realizations of quantum gates that are useful for quantum information processing. After a brief introduction into the subject, in Sect. 2 we focus on a particular way of using atoms in one-dimensional optical lattices as carriers of the quantum information. As an alternative, in Sect. 3 the information carriers are photons that interact via effective nonlinearities which arises from mixing at passive linear optical elements and post-selection through photodetection. These two seemingly different implementations have in common that their decoherence mechanism is described by a single theory, namely that of quantum electrodynamics in causal media which will be the subject of Sect. 4. Figure 1 should serve as an overview of the subject areas covered.

1 Introduction

1.1 Why Quantum Information Processing?

Apart from the academically driven curiosity to learn more about information in a quantum-mechanical setting, there is very urgent practical need

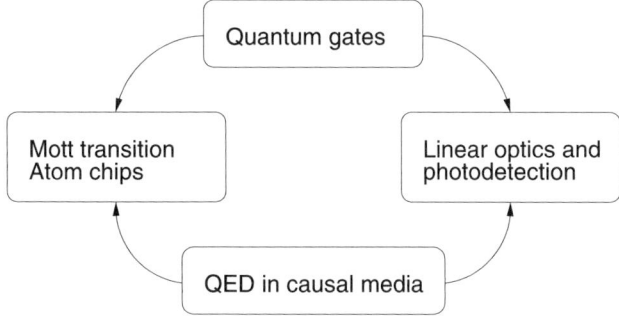

Fig. 1. Connections between the subject areas covered

S. Scheel et al.: *Quantum Gates and Decoherence*, Lect. Notes Phys. **689**, 47–81 (2006)
www.springerlink.com   © Springer-Verlag Berlin Heidelberg 2006

to investigate information processing and computing from a quantum theory point of view. Over the last 30 years the number of transistors on an integrated circuit, i.e. the complexity of the computer made up from those chips, doubles roughly every 18 months. This empirical behaviour is famously known as Moore's first law (after the co-founder of Intel Corp., G. Moore). An extrapolation reveals that by the year 2017 a bit of information will need to be encoded in a single atom. Even if Moore's law breaks down before we eventually come to the point that, at the current growth rate, by the year 2012 the dimensions of logical elements will be so small that quantum effects upon computation cannot be neglected any longer.[1] Thus, technological progress necessitates the study of what implications quantum theory has on computation.

Another reason for looking deeper into quantum computing lies in the potential ability to simulate the temporal evolution of (possibly chaotic) quantum systems. The intrinsic and potentially massive parallelism of quantum computers that originates in the linearity of quantum mechanics and the resulting superposition principle, would allow to investigate quantum Hamiltonian systems without the need for an exponential temporal overhead on a classical computer (see e.g. [1]).

1.2 Quantum Gates vs. Classical Gates

"Information is physical."
— Rolf Landauer

The most important difference between a logical gate, as known from classical information processing, and a quantum gate is its reversibility or conservation of information. In order to see what that means, let us consider for example the logical NAND gate whose action on logical zeros and ones are given by the truth Table 1. Equivalently, we can describe it by the ac-

Table 1. Truth table of the logical NAND gate

Inputs	Output
0 0	1
0 1	1
1 0	1
1 1	0

tion of $\overline{X_1 \wedge X_2}$ on two Boolean variables X_1 and X_2. This gate is generic

---

[1] The most recent Intel Pentium 4 processor already contains features of the size of 90 nm, i.e. 900 Å. The lithography used to generate such structures therefore uses XUV light!

for classical gates in that it has two (or more) inputs and only a single output. In quantum mechanics, we are allowed to form linear combinations or superpositions of the logical basis states, i.e. $c_0|00\rangle + c_1|01\rangle + c_2|10\rangle + c_3|11\rangle$. The formal application of the classical NAND on this quantum superposition would result in a state $\propto c_3|0\rangle + (c_0 + c_1 + c_2)|1\rangle$. The resulting state contains far less information than the original state. Whereas before the gate operation all the weight coefficients c_i gave us information about the quantum state, after the action of the classical NAND only the *sum* $c_0 + c_1 + c_2$ plays a rôle in determining the weight of the logical $|1\rangle$. All the information about the mutual *differences* are lost during the operation. However, this state is obviously not properly normalized anymore. The formal action of a classical gate is therefore actually an ill-posed operation in quantum mechanics. In fact, there is no way of making sense of classical logical operations acting on quantum superpositions.

In quantum information processing, all operations have to be unitary. This means on one hand that the number of output degrees of freedom must equal the number of input degrees of freedom, and on the other hand it means that no information is lost about an initial superposition. In general, we could allow even more general operations such as completely positive maps of which the unitary operations are a subset. However, equal number of inputs and outputs and conservation of information are not sufficient to define a quantum operation. For example, let us consider an operation in which a given quantum state $c_0|0\rangle + c_1|1\rangle$ is transformed into its orthogonal complement, hence into the state $-c_1^*|0\rangle + c_0^*|1\rangle$. This operation could be thought of as being the analogue of the classical NOT operation which is defined as $X \mapsto \overline{X} : \overline{X} \wedge X = 0$. However, it turns out that the quantum-NOT operation is not a completely positive map and therefore cannot be implemented by any quantum circuit. In fact, the map describing this transformation is anti-unitary [2] with determinant -1.[2]

Universal Set of Quantum Gates

Now that we have shown that there is no strict connection between classical logical gates and quantum mechanics, we will next briefly discuss which operations are compatible with quantum mechanics. Let us first look into operations acting on a single logical qubit state $c_0|0\rangle + c_1|1\rangle$. The dynamical group associated with this state is the unitary group SU(2) whose action can be given in terms of its generators, the Pauli spin operators $\hat{\sigma}_i$ ($i = x, y, z$), as $\hat{U} = e^{i\boldsymbol{\alpha} \cdot \hat{\boldsymbol{\sigma}}}$. Here, $\boldsymbol{\alpha}$ denotes the direction of the rotation axis and the associated rotation angle. By decomposing $\hat{U}$ into exponential factors, one can derive the so-called Euler decomposition $\hat{U} = e^{ia_1\hat{\sigma}_z} e^{ia_2\hat{\sigma}_y} e^{ia_3\hat{\sigma}_z}$ which just

---

[2]In this context, the notion of universal gate has been invented. A universal NOT-gate would be the one that, when averaging over all input states, comes closest with respect to some measure to the NOT-operation.

describes a sequence of elementary rotations. The Pauli operators have the following representation in the single-qubit basis:

$$\hat{\sigma}_x = |0\rangle\langle 1| + |1\rangle\langle 0| \,,$$
$$\hat{\sigma}_y = i(|0\rangle\langle 1| - |1\rangle\langle 0|) \,,$$
$$\hat{\sigma}_z = |0\rangle\langle 0| - |1\rangle\langle 1| \,. \tag{1}$$

An equivalent set of operators would be $\hat{\sigma}_\pm = \hat{\sigma}_x \pm i\hat{\sigma}_y$ and $\hat{\sigma}_z$ which results in an operator decomposition of the form $\hat{U} = e^{b_1(1+\hat{\sigma}_z)}e^{b_2\hat{\sigma}_+}\,e^{b_3\hat{\sigma}_-}e^{b_4(1-\hat{\sigma}_z)}$. This result will be used later in Sect. 3 for the description of beam splitters.

By definition, the Pauli operators are enough to represent all other single-qubit operations. An important example is the Hadamard gate $\hat{H}$ (not to be confused with the Hamiltonian),

$$\hat{H} = \frac{1}{\sqrt{2}}\Big[(|0\rangle + |1\rangle)\,\langle 0| + (|0\rangle - |1\rangle)\,\langle 1| \Big] \,, \tag{2}$$

which transform logical basis states into coherent superpositions. From the definition of the Pauli operators it is obvious that the Hadamard gate can be written as $\hat{H} = (\hat{\sigma}_x + \hat{\sigma}_z)/\sqrt{2}$.

As for two-qubit gates, the structure of a general unitary operator is not so obvious. We will therefore restrict our attention to some particularly useful gates. One of the most important nontrivial two-qubit gates is the controlled-phase gate $\hat{C}_\varphi$ whose truth table is given in Table 2. The controlled-phase

Table 2. Truth table of the controlled-phase gate

Input	Output
$|00\rangle$	$|00\rangle$
$|01\rangle$	$|01\rangle$
$|10\rangle$	$|10\rangle$
$|11\rangle$	$e^{i\varphi}|11\rangle$

gate (or its special case the controlled-$\hat{\sigma}_z$ gate with $\varphi = \pi$), together with the set of Pauli operators, forms a universal set of quantum gates [3] by which we mean that all possible quantum networks can be built up from them. Therefore, if one is able to build these four gates, one can generate arbitrary quantum operations by concatenating them (Fig. 2).

For example, another nontrivial two-qubit gate is the controlled-NOT (or CNOT) under whose action the logical states $|10\rangle$ and $|11\rangle$ are exchanged and the other left untouched.[3] This operator can be realized by the combination $\hat{H}_1\hat{C}_\pi\hat{H}_1$, i.e. by two Hadamard gates acting on mode 1 and a controlled-$\hat{\sigma}_z$ gate. Other two-qubit gates of importance are the swap in which the

---

[3]The associated operator can be written as $|0\rangle\langle 0| \otimes \hat{I} + |1\rangle\langle 1| \otimes \hat{\sigma}_z$.

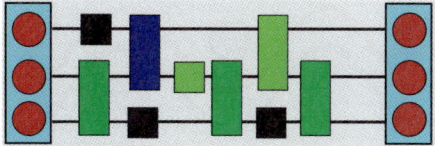

Fig. 2. A quantum network can be built from a universal set of quantum gates. Here we sketch a three bit register which is manipulated by a sequence of one and two qubit gates to realize a chosen unitary transformation which effects the relevant algorithm

qubits are simply exchanged and the square root of swap which serves as an entangling gate.

2 Atomic Realisation – Atom Chips and the Mott Transition in Optical Lattices

There are many possible realizations of quantum bits: laser-cooled trapped ions, cold trapped neutral atoms, atoms in high-Q single-mode cavities as well as condensed matter candidates such as quantum dots or Josephson junctions. For any one of these to be viable, the effects of the environment (dissipation and temperature) must be minimized, if not entirely eradicated. This simply means we need to use cold state initialization.

In this section, we will describe a possible way of implementing quantum computation with cold atom technology. This includes the application of optical lattices in a sufficiently cold cloud of atoms showing Bose–Einstein condensation (BEC). The lasers which generate the periodic spatially varying trapping potential through the AC Stark effect create an optical lattice strong enough to induce a quantum phase transition from the superfluid phase that characterises the BEC to the Mott insulator phase, characterised by a regular structure of one atom per lattice site that can serve as a quantum register. A universal set of quantum gates can then be realised by manipulations of the lattice potential with additional laser fields.

2.1 Bose–Einstein Condensates and the Mott Transition

In recent years it has been realized that Bose–Einstein condensates (BECs for short) can undergo a phase transition if loaded into a three-dimensional periodic potential which for example can be realized by standing-wave optical fields [4]. That is, one starts off with a BEC in its superfluid phase in which the relative phases (or rather correlations) between the atoms are well-defined such that the whole ensemble of atoms can be described by a single macroscopic wave function (in first approximation). By loading this conden-

Fig. 3. Counter-propagating laser beams induce a periodic spatially varying trapping potential through the AC Stark shift

sate into a weak optical lattice (see Fig. 3) the number of atoms per lattice site is undetermined and can vary widely. The ground-state wave function for N atoms in a lattice with M sites is therefore

$$|\Psi_S\rangle \propto \left(\sum_{i=1}^{M} \hat{a}_i^\dagger\right)^N |0\rangle \tag{3}$$

where the $\hat{a}_i^\dagger$ denote creation operators of an atom at the lattice site i.

However, when increasing the strength of the potential by increasing the power of the laser beams that create the standing-wave potential, eventually there will be a phase-transition to a state of the condensate in which each lattice site is occupied by a fixed and well-defined number of atoms (ideally we would like to have exactly one atom per site). In this so-called Mott-insulator phase the relative phases (or correlations) between neighboring lattice sites are undetermined. The ground-state wave function here looks essentially like

$$|\Psi_M\rangle \propto \prod_{i=1}^{M} (\hat{a}_i^\dagger)^n |0\rangle \tag{4}$$

where n is the number of atoms per lattice site. Experimental evidence of this phase-transition has been obtained in the beautiful experiments described in [5, 6]. Although a Bose–Einstein condensate really exists only in three dimensions (since only there we find a phase transition from a thermal cloud to a condensate), there are analogous systems such as the quasi-condensate [7] and the Tonks–Girardeau gas [8] in one dimension that have similar properties.

The effective interaction Hamiltonian that can be derived from the Gross–Pitaevskii equation under the assumption that the atoms have localized single-particle wave functions can be written as

$$\hat{H} = \frac{U}{2}\sum_i \hat{n}_i(\hat{n}_i - 1) - J\sum_i (\hat{a}_i^\dagger \hat{a}_{i+1} + \hat{a}_{i+1}^\dagger \hat{a}_i) \,. \tag{5}$$

This Hamiltonian is also known in the literature as the Bose–Hubbard Hamiltonian. The first term with coupling strength U is the collisional energy of atoms occupying the same lattice site. Obviously, if there is zero or just one atom per site, this term vanishes identically. The second term is the so-called hopping term which essentially is given by the overlap of the localized wave functions at neighboring sites and describes tunneling between adjacent lattice sites with tunneling strength J. If, for example, the optical lattice is formed by the standing wave of a one-dimensional cavity mode having width L, the trapping potential is given by

$$V(x) = -V_0 \sin^2 kx \exp\left(-\frac{2r^2}{L^2}\right) \tag{6}$$

where r denotes the transverse distance from the lattice axis and k the wave number of the standing wave. With this potential, the collisional strength can be approximated by

$$U \approx \frac{4a_s V_0^{3/4} E_R^{1/4}}{\sqrt{\lambda L}} \tag{7}$$

where a_s is the scattering length of the atomic collisions and E_R the atomic recoil energy. A derivation shows that the tunneling rate can be expressed as

$$J \approx \frac{E_R}{2} \exp\left(-\frac{\pi^2}{4}\sqrt{\frac{V_0}{E_R}}\right)\left[\sqrt{\frac{V_0}{E_R}} + \left(\sqrt{\frac{V_0}{E_R}}\right)^3\right]. \tag{8}$$

It is now apparent that by changing the potential depth V_0 one can tune the system of atoms into either one of the two phases (superfluid or Mott-insulator) as seen in Fig. 4. For example, if V_0 is relatively small compared to the recoil energy E_R, the tunneling rate J will be large and the atoms will be delocalized and form a superfluid. Increasing V_0 means exponentially decreasing the tunneling rate and the atoms will become stuck in their respective potential wells and form the Mott-insulator phase. In Fig. 5 we show an example calculation for the population distribution

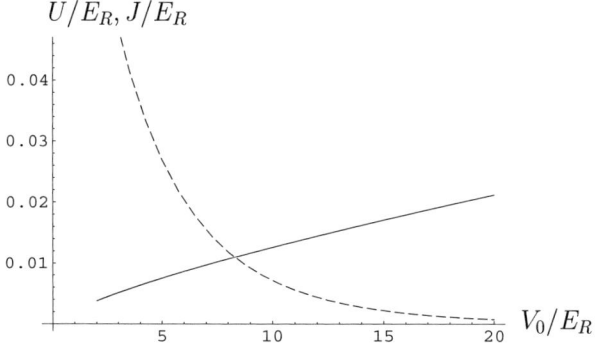

Fig. 4. The values of U/E_R (*solid line*) and J/E_R (*dashed line*) are plotted against V_0/E_R for typical values of the relevant length scales appearing in (7) ($a_s = 5.6\,\text{nm}$ for ^{87}Rb, $L = \lambda = 10\,\mu\text{m}$)

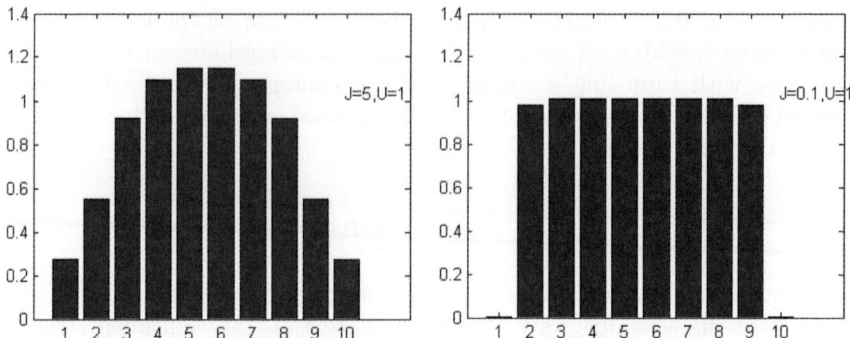

Fig. 5. Distribution of 8 atoms among 10 lattice sites in the limits $U/J \ll 1$ (*left figure*) and $U/J \gg 1$ (*right figure*)

in a one-dimensional lattice with 10 sites occupied by 8 atoms. The figure on the left depicts a situation in which the collisional energy U is small compared to the tunneling rate J ($U/J \ll 1$) and hence a superfluid phase exists, witnessed by a Gaussian-like distribution. On the other hand, the figure on the right shows the one-by-one distribution of atoms for $U/J \gg 1$ which closely resembles a Fock state at each site. In fact, the critical value for obtaining the phase-transition is $U/J \approx 11.6$ [9]. Note that the calculation of the ground-state wave function can only be done numerically, which amounts to computing the eigenvector corresponding to the smallest eigenvalue of a sparse matrix of dimension $\binom{A+W-1}{A}$, where A is the number of atoms in W lattice sites.

2.2 Quantum Computation with a 1D Optical Lattice

In order to use the Mott-insulator phase for building a quantum register, it is advantageous to use atoms with two degenerate ground states $|g_a\rangle$ and $|g_b\rangle$ which are coupled to each other with a Raman transition via an excited state [10]. In order to trap both "species" of atoms simultaneously, an optical lattice formed of two counterpropagating laser beams with perpendicular linear polarization vectors is needed. The result of this configuration is that the atoms are trapped in two overlapping optical lattices with polarizations σ_+ and σ_- each of which can trap one of the two atomic ground states.

If we denote the creation operators of atoms in the states $|g_a\rangle$ and $|g_b\rangle$ by $\hat{a}^\dagger$ and $\hat{b}^\dagger$, respectively, we can write the interaction Hamiltonian (5) as

$$\hat{H} = \sum_i \left[\frac{U_{aa}}{2} \hat{n}_i^a (\hat{n}_i^a - 1) + U_{ab} \hat{n}_i^a \hat{n}_i^b + \frac{U_{bb}}{2} \hat{n}_i^b (\hat{n}_i^b - 1) \right]$$
$$- \sum_i \left(J_i^a \hat{a}_i^\dagger \hat{a}_{i+1} + J_i^b \hat{b}_i^\dagger \hat{b}_{i+1} + J_i^R \hat{a}_i^\dagger \hat{b}_i + \text{h.c.} \right). \tag{9}$$

Here we encounter several couplings amongst atoms of the same species (collisional couplings U_{aa}, U_{bb} and tunneling rates J^a, J^b) as well as coupling of atoms of different species (collisional coupling U_{ab} and effective Raman coupling J^R).

We assume that initially all atoms are in the ground state $|g_a\rangle$. This state will be denoted by $|0\rangle$, whereas the second ground state $|g_b\rangle$ will serve as the logical $|1\rangle$.

Single-Qubit Rotations

From the above definition of the logical states it is clear that single-qubit rotations can be performed by a Raman process. For example, starting at $|0\rangle$ and performing half a Raman cycle leaves us with an atom in state $|1\rangle$, whereas shorter interactions produce superposition states between $|0\rangle$ and $|1\rangle$. If one starts from an arbitrary superposition, the application of a full Raman cycle generates an effective $\hat{\sigma}_x$ rotation. In addition, a simple Rabi cycle of 2π applied to the ground state $|g_b\rangle$ is equivalent to changing the phase of the logical $|1\rangle$ by π and thus represents a σ_z operation.

Two-Qubit Gates – Controlled-Phase Gate and Square Root of the Swap Operator

In order to realize two-qubit operations, we need to couple two atoms in neighboring lattice sites to each other in a controlled way. Let us assume that the lattice is sufficiently deep such that, without external changes, tunneling between lattice sites is prohibited, hence J^a and J^b can be initially neglected. A tunneling coupling between neighboring sites can be switched on by applying an additional standing wave perpendicular to the lattice with a waist that should not exceed the size of two lattice sites (see Fig. 6). Depending on the circular polarization (σ_+ or σ_-) of the additional laser beam the tunneling strength for both atomic species can be tuned independently.

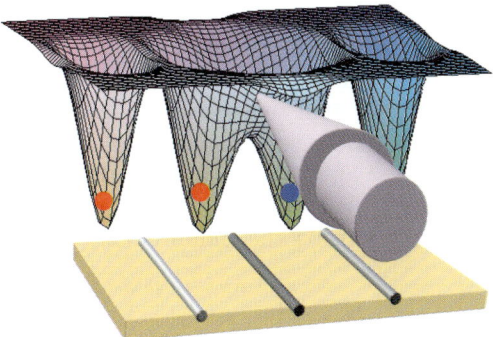

Fig. 6. The tunneling interaction between neighboring lattice sites can be switched on by lowering the potential barrier between the sites

For the two-qubit operations to work, the tunneling coupling must always be much smaller than the collisional coupling, even when the potential barrier has been lowered. Hence, we require $U \gg J$ for all times during the interaction process. This amounts to the fact that the system stays in the eigenspace of the collisional terms U_{aa} and U_{bb}, hence in a number state. This eigenspace is in fact degenerate with respect to the occupation numbers $n_{a,b} = 0$ or $n_{a,b} = 1$ since the collisional energy is obviously zero in both cases. There is an energy gap from this degenerate subspace to the state with two atoms of the same species per lattice site which means that those states can be adiabatically eliminated from the evolution. If, in addition, there is a large inter-species collisional coupling U_{ab}, then the same energy gap persists even for states in which two atoms of different species occupy one lattice site. Thus, every state with more than one atom per lattice site, regardless of their species, can be adiabatically eliminated, thereby leaving us with a degenerate eigenspace spanned by the logical states $|0\rangle = |n_a = 1, n_b = 0\rangle$ and $|1\rangle = |n_a = 0, n_b = 1\rangle$. This constitutes our well-defined computational space.

In what follows, we will use the notation introduced in [10], denoting the state of the atomic population in two lattice sites by $|n_a^1, n_b^1; n_a^2, n_b^2\rangle$. Then, for example, a state of two atoms in their respective ground states $|g_b\rangle$ is given by $|01; 01\rangle$ and represents the logical two-qubit state $|11\rangle$. Suppose now we were to lower the potential barrier between two neighboring sites only for the atom in ground state $|g_b\rangle$, which can be done by choosing an appropriate polarization of the incident laser beam. Then, according to the Hamiltonian (9), this state can couple to only two other states, $|02; 00\rangle$ and $|00; 02\rangle$, with two atoms simultaneously at one lattice site. The Hamiltonian (9) can be written in the basis $\{|01; 01\rangle, |02; 00\rangle, |00; 02\rangle\}$ as

$$H_{bb} = \begin{pmatrix} 0 & -J^b & -J^b \\ -J^b & U_{bb} & 0 \\ -J^b & 0 & U_{bb} \end{pmatrix}. \tag{10}$$

This Hamiltonian effectively corresponds to a V-system with ground state $|01; 01\rangle$ and excited states $|02; 00\rangle, |00; 02\rangle$ coupled by an effective Rabi frequency $-J^b/2$ and detuned by U_{bb} (see left figure in Fig. 7). Assuming that the detuning is large, the system remains in the ground state and acquires a phase $-2 \int dt \, (J^b)^2/U_{bb}$. However, in the same manner the logical states $|01\rangle = |1, 0; 0, 1\rangle$ and $|10\rangle = |0, 1; 1, 0\rangle$ acquire phases. This is due to their interaction with the states $|1, 1; 0, 0\rangle$ and $|0, 0; 1, 1\rangle$, respectively. For example, in the basis $\{|1, 0; 0, 1\rangle, |1, 1; 0, 0\rangle\}$ the Hamiltonian (9) simply reads

$$H_{ab} = \begin{pmatrix} 0 & -J^b \\ -J^b & U_{ab} \end{pmatrix} \tag{11}$$

which leads to a phase shift of $\int dt \, (J^b)^2/U_{ab}$. The latter can be compensated for by applying a single-qubit rotation on the state $|1\rangle$ on both lattice sites.

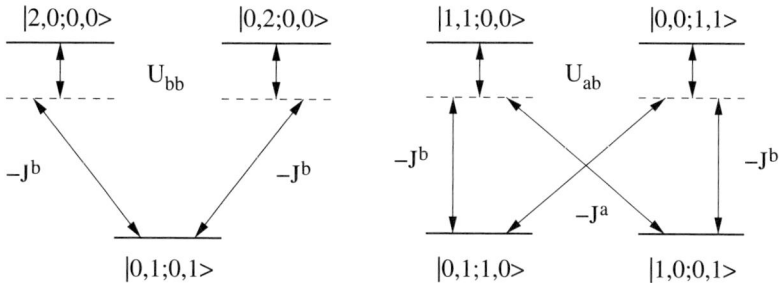

Fig. 7. Effective 3-level system coupling the state $|0, 1; 0, 1\rangle$ to its excited states (*left figure*), and effective 4-level system coupling $|0, 1; 1, 0\rangle$ and $|1, 0; 0, 1\rangle$ to their respective excited states (*right figure*)

The overall effect is to introduce a phase shift

$$\varphi = 2 \int_0^T dt \left(\frac{(J^b)^2}{U_{ab}} - \frac{(J^b)^2}{U_{bb}} \right) \tag{12}$$

to the logical state $|11\rangle$ while keeping all other logical basis states unchanged. This is the controlled-phase gate which, together with the single-qubit rotations, constitutes a universal set of operations for quantum computing.

Similarly, we can choose to act upon the logical states $|01\rangle$ and $|10\rangle$ which contain one atom of each species in neighboring lattice sites. In the adiabatic approximation, the Hamiltonian connecting these two states is (in the basis $\{|1, 1; 0, 0\rangle, |1, 0; 0, 1\rangle, |0, 1; 1, 0\rangle, |0, 0; 1, 1\rangle\}$, see right figure in Fig. 7)

$$H_{ab} = \begin{pmatrix} U_{ab} & -J^a & -J^b & 0 \\ -J^a & 0 & 0 & -J^b \\ -J^b & 0 & 0 & -J^a \\ 0 & -J^b & -J^a & U_{ab} \end{pmatrix} \tag{13}$$

with the solution that the following transformation on the logical states $|01\rangle$ and $|10\rangle$ is achieved:

$$|01\rangle \mapsto e^{-i\varphi} [\cos I |01\rangle - i \sin I |10\rangle]$$
$$|10\rangle \mapsto e^{-i\varphi} [-i \sin I |01\rangle + \cos I |10\rangle] \tag{14}$$

with

$$\varphi = \int_0^T dt \frac{(J^a)^2 + (J^b)^2}{U_{ab}}, \quad I = 2 \int_0^T dt \frac{J^a J^b}{U_{ab}}. \tag{15}$$

The effective Hamiltonian is therefore just

$$H_{\text{eff}} = -I| (|10\rangle\langle 01| + |01\rangle\langle 10|), \tag{16}$$

which, for $I = \pi/4$ is also known also the square root of the swap operator. Additionally, the logical states $|00\rangle$ and $|11\rangle$ acquire phases $-2 \int dt \, (J^a)^2/U_{aa}$ and $-2 \int dt \, (J^b)^2/U_{bb}$, respectively, as discussed before. Thus, the two qubits pick up only an *overall* phase if one chooses $(J^a)^2/U_{aa} + (J^b)^2/U_{bb} = [(J^a)^2 + (J^b)^2]/U_{ab}$. This overall phase can later be reversed by applying appropriate single-qubit rotations on both lattices sites.

These examples show that it should be straightforward, in principle, to implement different types of single- and two-qubit operations with very few laser pulses (typically just one). The drawback is that for all above considerations the adiabaticity condition must be fulfiled. Thus, the couplings J^a and J^b are assumed to be small compared to the collisional couplings U_{aa}, U_{bb}, and U_{ab} which results in unwanted long gate evolution times. For example, in [10] it has been estimated that, for currently measured collisional couplings of $\mathcal{O}(1\,\text{kHz})$ [11], the gate operation time for nontrivial gates such as the above-described controlled-phase gate with an error rate less than 10^{-3} is roughly $100\,\text{ms}$. This time scale is far too long to render quantum computation useful with this scheme. A way to circumvent this problem and to drastically reduce gate operation times is to relax the adiabaticity condition and to note that even without adiabatic evolution there are certain instances in which the atomic population returns completely to the logical space in which we have started. The drawback here is that laser amplitudes and pulse durations have to be stabilized much more precisely than in the adiabatic regime. Despite that, it seems that there is much potential in this and other proposals for quantum computing on an optical lattice.

Moreover, this time scale is already of the order of the currently possible trapping lifetime of atoms in recent experiments (see Sect. 4 for a detailed discussion about trapping losses). Hence, only a few gate operations can be performed before the atoms are lost from the optical lattice.

2.3 Experimental Realization with Atom Chips

A particularly interesting way of implementing the above ideas of atomic registers is by using atom chips (for reviews on this exciting subject, see for example [12, 13]). The basic idea here is to trap cold atoms in one of their low-field seeking hyperfine ground states in the combined magnetic fields of a current-carrying wire and a constant transverse bias (see Fig. 8). The radial magnetic field of the wire is superimposed on a constant homogeneous field from the side thereby creating a line of zero magnetic field parallel to the wire. The atoms are thus trapped in a tubular region whose distance from the wire can be tuned by adjusting the relative strengths of the two overlapping magnetic fields. However, as depicted in the inset of Fig. 8, the atoms can flip their magnetic sublevels once they reach the region of zero magnetic field, hence yet another bias field has to be applied, this time parallel to the wire, to prevent the atoms from doing that.

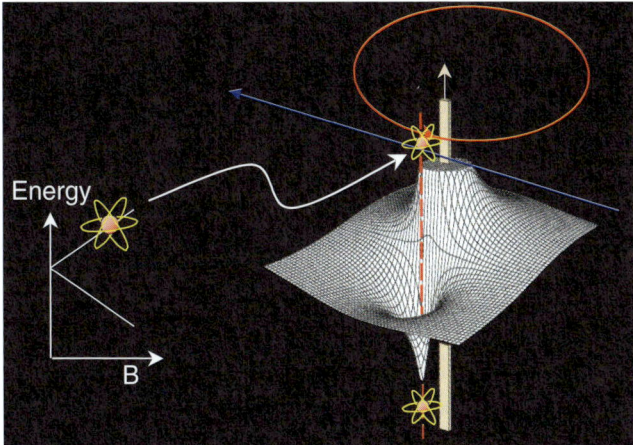

Fig. 8. Principle of the magnetic guide for atoms (see text for details)

Once the atoms are trapped and evaporatively cooled to form a Bose–Einstein condensate, the atom cloud has typically a length of 100 μm and a diameter of 2 μm, hence it forms a long and thin cigar-shaped object which can be treated as quasi-1D. The next step is to confine the cloud between two highly reflecting mirrors that form a microcavity in the standing wave of which the atoms experience the periodic light potential we were envisaging earlier (Fig. 9). The idea is then to move a string of atoms [14] in a condensate state into the cavity region to perform the transition to the Mott-insulator. The current experimental status is that experiments have been performed in which a cloud of trapped BEC has been moved in a "conveyor-belt" fashion across the surface of a miniaturized atom chip. The next step will be to

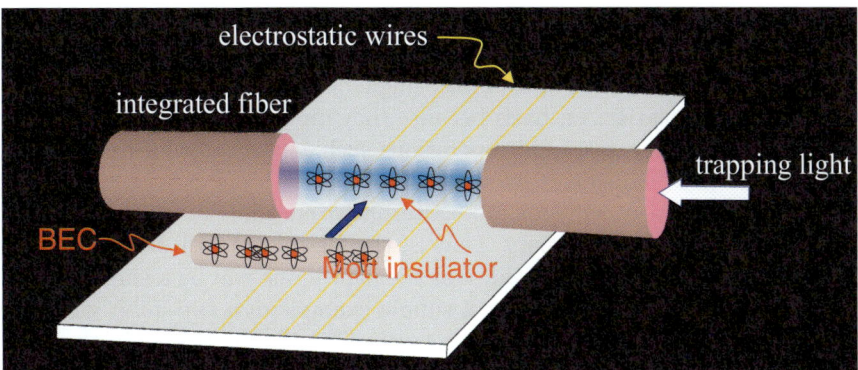

Fig. 9. Proposed scheme for combining atom chips with optical cavities for generating one-dimensional optical lattices

integrate two adjacent optical fibres with focusing ends onto the atom chip to create the standing-wave potential.

An obvious performance limitation is given by the time the atoms actually spend in their respective trapped states above the wire. Several noise sources, both technical and fundamental, can cause spin flips from a trapped to an anti-trapped magnetic sublevel that will prevent the atoms from staying indefinitely above the wire surface. The influence of the predominant noise source – magnetic fluctuations caused by absorption in the current-carrying wire – will be investigated in more detail in Sect. 4.

3 Photonic Realisation – Passive Linear Optics and Projective Measurements

In a rather different setting compared to Sect. 2 we can consider photons as the carriers of quantum information. Photons constitute an alternative to atoms – being qubits at rest – as they are massless particles and therefore move at the speed of light which earned them the name "flying" qubits. They will eventually be part of larger networks and are considered to be vital in transporting quantum information over longer distances. Moreover, it is believed that photons could even be used to perform quantum operations themselves. It is therefore vital to know what kind of operations can be done with photons and how they are implemented.

3.1 Qubit Encoding and Single-Qubit Operations

First of all, it is necessary to define which photonic degrees of freedom we would like to encode our qubits in. One possibility would be to encode the information in the polarization state (horizontal or vertical) [15, 16], another one the superposition states of one photonic excitation in two modes [17]. Another, seemingly complementary encoding would use the photon number or Fock states of a photon. All the mentioned possibilities have their advantages and disadvantages. For example, single-qubit rotations are easily implemented in the polarization basis because they are just performed by $\lambda/4$- or $\lambda/2$-plates. However, two-qubit operations such as the controlled-phase gate are impossible to implement by wave plates.

In what follows we will look at an alternative encoding in which the qubits are defined by photon numbers. That is, the logical zero will be the vacuum state of the electromagnetic field, and the logical one will be a single-photon Fock state. Then, let us consider a simple example of a two-qubit gate, the controlled-phase gate defined by the truth Table 2. That is, only the basis state containing one photon in each mode will pick up a phase φ, all other basis states are left unchanged.

Here we immediately encounter a problem in that, in order to realize this quantum gate, two single photons have to interact with each other sufficiently strongly to produce the desired phase shift. In fact, nature is not so kind as to allow us to do this easily. Consider for a moment standard quantum electrodynamics. The lowest-order Feynman diagram that contains a photon-photon interaction is depicted in Fig. 10. The interaction strength is of second order in the fine structure constant $\alpha \approx 1/137$ and therefore negligible.

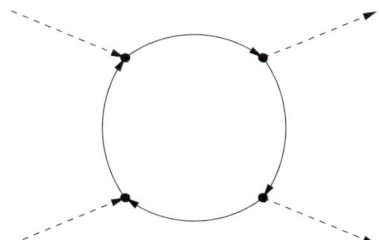

Fig. 10. Lowest-order Feynman diagram leading to a photon-photon interaction

A seemingly promising alternative is provided by nonlinear materials exhibiting a $\chi^{(3)}$ or Kerr nonlinearity. The constituents of those materials respond nonlinearly to an external electromagnetic field, and after tracing out the matter degrees of freedom leave behind an effective nonlinear interaction between photons. However, these natural nonlinearities are still too small to be of any use since only phase shifts of the order of 10^{-8} can be achieved.

3.2 Measurement-Induced Nonlinearities

A possible way out of this dilemma can be found in the use of so-called measurement-induced nonlinearities. The idea is rather simple. Suppose we wanted to act on a pure single-mode state $|\psi\rangle$ with a nonlinear operator. In order to achieve that, we mix our signal state with another pure single-photon state at a beam splitter and perform a suitable projection measurement at one output port of the beam splitter. The result will be a *conditional* nonlinear operator acting on the signal mode. For example, let us consider the situation depicted in Fig. 11 in which a single-photon Fock state acts as our auxiliary (or ancilla) state. It is well known that a beam splitter acts as an SU(2) group element on the photonic amplitude operators [18]. Hence, the operator

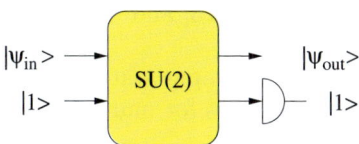

Fig. 11. Single beam splitter acting conditionally as a nonlinear operator

associated with the action of the beam splitter can be written in the two-mode representation of the Pauli operators (see Sect. 1) as [19]

$$\hat{U}_{12} = T^{\hat{n}_1} e^{-R^* \hat{a}_2^\dagger \hat{a}_1} e^{R \hat{a}_1^\dagger \hat{a}_2} T^{-\hat{n}_2} \tag{17}$$

where $1, 2$ label the modes impinging on the beam splitter. Consider now the situation in which the auxiliary mode is prepared in a single-mode Fock state and a single photon is detected in the same output port. Then the resulting *conditional* operator acting on the signal mode reads

$$\hat{Y}_1 = \langle 1_2 | \hat{U}_{12} | 1_2 \rangle = T^{\hat{n}_1 - 1} \left[|T|^2 - \hat{n}_1 |R|^2 \right] . \tag{18}$$

It is clear from the expression (18) that this operator represents a nonlinear evolution since it is inherently quartic in the photonic amplitude operators $\hat{a}_1^{(\dagger)}$. This is an *effective* nonlinearity since the overall evolution, when taking into account the contributions from all possible measurement outcomes, is still perfectly *linear*. One observes furthermore that the operator $\hat{Y}_1$ is independent of the signal state, in fact, the nonlinearity is created by the measurement process only. Therefore, following the notation commonly used in quantum optics, this represents quantum-gate engineering – as opposed to quantum-state engineering – or, in other words, quantum-state engineering of arbitrary states. Note also that in general the success probability $p = \|\hat{Y}_1|\psi\rangle\|$ does depend on the chosen signal state $|\psi\rangle$. However, if the beam splitter parameters are chosen such that $\hat{Y}_1^\dagger \hat{Y}_1 |\psi\rangle \propto |\psi\rangle$, i.e. the conditional operator is proportional to a unitary operator (hence a quantum gate), then the success probability is state-independent as well. This important fact will be used later on to actually construct quantum gates with maximal success probability.

In order to go one step further in the development of conditional quantum gates for photons, we have to look at larger beam splitter networks consuming more than just one auxiliary or ancilla state. To do so, we note that every U(N) transformation of N photonic amplitude operators can always be realized by a triangular-shaped network of at most $N(N-1)/2$ beam splitters and some phase shifters [21]. Hence, we need to generalize our arguments to whole networks of beam splitters. There seem to be essentially two ways of analyzing U(N) networks. One is to look at the unitary operator $\hat{U}_{12...N}$ acting on N modes in terms of its Euler decomposition analogous to (17). Although this is not completely impossible, it is very tedious indeed. In fact, the Euler decomposition has only been calculated for $N = 3, 4$ [22]. An alternative way has been developed in [23] and uses well-known techniques from bosonic operator algebras.

For this purpose, we write the input state – still assumed to be a single-mode state, the theory is analogous for multi-mode states – in a functional form as

$$|\psi\rangle = \hat{f}(\hat{a}_1^\dagger)|0\rangle = \sum_m \frac{c_m}{\sqrt{m!}} (\hat{a}_1^\dagger)^m |0\rangle , \tag{19}$$

where the c_m are constrained in such a way that $\sum_m |c_m|^2 = 1$. Analogously, we write the auxiliary state $|A\rangle$ and the state $|P\rangle$, respectively, in product form as

$$|A\rangle = \prod_{i=2}^{N} \frac{(\hat{a}_i^\dagger)^{m_i}}{\sqrt{m_i!}} |0\rangle^{\otimes N-1}, \quad |P\rangle = \prod_{j=2}^{N} \frac{(\hat{a}_j^\dagger)^{n_j}}{\sqrt{n_j!}} |0\rangle^{\otimes N-1}. \tag{20}$$

Here m_i is a non-negative integer that represents the number of photons initially in mode i, and n_j is the number of photons in the projected mode j. The beam splitter network is represented by a unitary $N \times N$-matrix $\boldsymbol{\Lambda}$, under the action of which the amplitude operators transform as

$$\hat{\mathbf{a}} \mapsto \boldsymbol{\Lambda}^+ \hat{\mathbf{a}}, \quad \hat{\mathbf{a}}^\dagger \mapsto \boldsymbol{\Lambda}^T \hat{\mathbf{a}}^\dagger. \tag{21}$$

Combining all these definitions we derive the (un-normalised) output state after mixing at the beam splitter network and projecting onto $|P\rangle$ as

$$|\psi'\rangle \propto \langle P | \hat{U}_{12...N} | A\rangle \otimes |\psi\rangle$$

$$= {}^{N-1\otimes}\langle 0| \prod_{i,j=2}^{N} \frac{(\hat{a}_j)^{n_j}}{\sqrt{m_i! n_j!}} \left(\sum_{k=1}^{N} \Lambda_{ki} \hat{a}_k^\dagger \right)^{m_i} \hat{f} \left(\sum_{l=1}^{N} \Lambda_{l1} \hat{a}_l^\dagger \right) |0\rangle^{\otimes N}. \tag{22}$$

We can see immediately from (22) that the effect of the beam splitter network is to mix the photonic creation operators of signal and auxiliary modes. At this point we use a well-known ordering formula

$$\left[\hat{a}, \hat{F}(\hat{a}, \hat{a}^\dagger) \right] = \frac{\partial}{\partial \hat{a}^\dagger} \hat{F}(\hat{a}, \hat{a}^\dagger) \tag{23}$$

to rewrite (22) in the convenient form

$$|\psi'\rangle \propto {}^{N-1\otimes}\langle 0| \prod_{i,j=2}^{N} \frac{\left(\frac{\partial}{\partial \hat{a}_j^\dagger} \right)^{n_j}}{\sqrt{m_i! n_j!}} \left(\sum_{k=1}^{N} \Lambda_{ki} \hat{a}_k^\dagger \right)^{m_i} \hat{f} \left(\sum_{l=1}^{N} \Lambda_{l1} \hat{a}_l^\dagger \right) |0\rangle^{\otimes N}. \tag{24}$$

In the following we quote some basic results that follow immediately from (24). Suppose all $N - 1$ auxiliary modes are prepared in single-photon Fock states, and all $N - 1$ photodetectors find only the vacuum state at the respective output ports. Then the output is proportional to the $N - 1$-fold application of the creation operator $(\hat{a}_1^\dagger)^{N-1}$. Similarly, if all auxiliary modes are prepared in the vacuum state and all photodetectors find exactly one photon each, then the output is proportional to $\hat{a}_1^{N-1}$. These results should not be surprising if one remembers that we are merely selecting a particular measurement result from a passive linear operation which preserves photon numbers. Somewhat more interesting is the situation in which all auxiliary modes are occupied by single photons and all detectors register a photon each. Then the associated conditional operator is a polynomial of degree $N - 1$ in the number operator $\hat{n}_1$, $P_{N-1}(\hat{n}_1)$. The last result is particularly important since it allows us to act upon the (unknown) coefficients of an N-dimensional signal mode *independently*.

3.3 Construction of Simple Quantum Gates

"The permanent doesn't really interact well with linear algebra."
— Mark Jerrum, private communication

The theory outlined above presents the framework which we will use to construct particularly simple and useful quantum gates. For simplicity, we first restrict ourselves to single-mode gates and give an outlook on multi-mode gates later. We also assume that we want to operate within Fock layers, i.e. we merely change phases of some expansion coefficients. The simplest nontrivial operations can be generated with signal modes that contain up to two photons.[4] Let us therefore consider the nonlinear phase shift gate $\hat{C}_\varphi$ which is defined by its action on a three-dimensional single-mode state as

$$c_0|0\rangle + c_1|1\rangle + c_2|2\rangle \overset{\hat{C}_\varphi}{\mapsto} c_0|0\rangle + c_1|1\rangle + e^{i\varphi}c_2|2\rangle. \tag{25}$$

According to what we have said earlier, in order to be able to act upon the coefficients of this three-dimensional signal state independently, we need a second-order polynomial in the number operator. Hence, we immediately find a network that performs the sought gate operation. The result is shown in Fig. 12. Note the triangular-shaped arrangement of the beam splitters as predicted in [21].

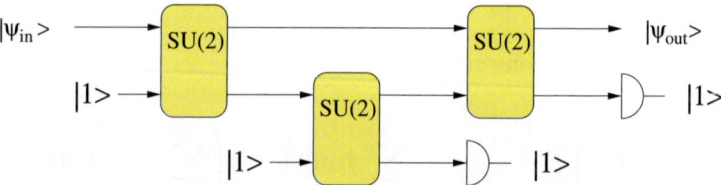

Fig. 12. SU(3) network realizing the nonlinear phase shift

Let us now analyze this network along the lines described above. We denote by $\boldsymbol{\Lambda}$ the unitary (3×3)-matrix associated with the SU(3) network. The conditional operator $\hat{Y}$ acting on the signal mode is then

$$\hat{Y}_1|\psi\rangle = c_0\mathrm{per}\,\boldsymbol{\Lambda}(1|1)|0\rangle + c_1\mathrm{per}\,\boldsymbol{\Lambda}|1\rangle$$
$$+c_2(2\Lambda_{11}\mathrm{per}\,\boldsymbol{\Lambda} - \Lambda_{11}^2\mathrm{per}\,\boldsymbol{\Lambda}(1|1) + 2\Lambda_{12}\Lambda_{21}\Lambda_{13}\Lambda_{31})|2\rangle \tag{26}$$

where per $\boldsymbol{\Lambda}$ denotes the permanent of the matrix $\boldsymbol{\Lambda}$ and per $\boldsymbol{\Lambda}(1|1)$ the permanent of $\boldsymbol{\Lambda}$ with the first row and column deleted. In order to realize a nonlinear phase shift, we need to fulfil the relations

---

[4]Note that all operations of the type $c_0|0\rangle + c_1|1\rangle \mapsto c_0|0\rangle + e^{i\varphi}c_1|1\rangle$ can be realized deterministically, since they are merely phase shifts.

$$\text{per } \boldsymbol{\Lambda}(1|1) = \text{per } \boldsymbol{\Lambda}, \quad \text{per } \boldsymbol{\Lambda}(1|1) \left[e^{i\varphi} + \Lambda_{11}^2 - 2\Lambda_{11} \right] = 2\Lambda_{12}\Lambda_{21}\Lambda_{13}\Lambda_{31} . \tag{27}$$

It turns out that there are infinitely many solutions to these equations which differ in their respective success probabilities. We have found numerically that the maximal probability is $1/4$ [23].

Permanents

The appearance of permanents in these problems is generic. The permanent of an $(n \times n)$ matrix $\boldsymbol{\Lambda}$ is a generalized matrix function, defined as

$$\text{per } \boldsymbol{\Lambda} = \sum_{\{\sigma_i\} \in S_n} \prod_{i=1}^{n} \Lambda_{i\sigma_i} \tag{28}$$

where S_n is the group of cyclic permutations [24]. Permanents naturally appear in combinatorial problems, graph theory and related subjects. In our context they "count" the ways of redistributing N single photons through an $SU(N)$ network to yield exactly N single photons at the outputs, i.e.

$$\text{per } \boldsymbol{\Lambda} = {}^{N\otimes}\langle 1 | \hat{U}_{12...N} | 1 \rangle^{\otimes N} . \tag{29}$$

As one can see from (26), the unitary network has to be adjusted such that the permanent and certain subpermanents of the associated unitary matrix fulfil certain relations. Furthermore, the overall success probability is just $|\text{per } \boldsymbol{\Lambda}(1|1)|^2$. Therefore, the optimal network is obtained by maximizing some permanent under a number of given constraints.

This is, in fact, a nontrivial task which is partly due to the fact that the algebraic property of a matrix being unitary does not imply any major simplifications for computing permanents.[5] One of the few known fact is that permanents of unitary matrices are bounded from above. This is a consequence of the Marcus–Newman theorem [24] which states that for all $(m \times n)$-matrices $\mathbf{U}$ and $(n \times m)$-matrices $\mathbf{V}$ the inequality

$$|\text{per } \mathbf{UV}|^2 \leq \text{per } \mathbf{UU}^* \text{per } \mathbf{VV}^* \tag{30}$$

holds which reduces, when setting $\mathbf{V} = \mathbf{I}$ and regarding $\mathbf{U}$ as being unitary, to $|\text{per } \mathbf{U}| \leq 1$. This is, of course, what one suspects if the permanent is supposed to be related to a success probability.

Another interesting fact to note is that computing permanents of matrices of increasingly larger size is a computationally hard problem in the sense that the computing time scales exponentially with the size of the matrix. Computer scientists say this problem is #P-complete. It seems that this computational problem is related to our quest to design quantum networks

---

[5]This is in stark contrast to the determinant which is just the product of the eigenvalues of a matrix.

that would eventually be able to solve problems in polynomial rather than exponential time. That is, we have to solve an exponentially hard problem *first* when designing networks in order to avoid it *later* when applying them to unknown quantum states in the course of a quantum computation. Note however, that this argument strictly applies only if one is interested in the optimal network. For most practical purposes it is sufficient to approximate the permanent which can be done efficiently with a quadratic number of steps.

Dimension of the Auxiliary State

What we have said so far about the nonlinear phase shift gate concerned the explicit construction of conditional nonlinear operators. In turns out, however, that this way of designing quantum gates, although being algorithmically transparent, does not yield optimal networks in terms of resources. For example, there is an alternative network proposed in Ref. [25] that also realizes the nonlinear phase shift but with one fewer beam splitter (see Fig. 13). As we will see later, the best networks for multi-dimensional quantum gates in terms of their respective success probabilities are not derived in this manner.

What seems to influence the required resources most is the dimensionality of the used auxiliary state. Comparing Fig. 12 and Fig. 13 one realizes that the Hilbert spaces spanned by the auxiliary modes are different in both cases. Starting from a product state of the form $|11\rangle$ as in Fig. 12, by beam splittings we actually span a three-dimensional space formed of the basis states $\{|20\rangle, |11\rangle, |02\rangle\}$. On the contrary, with an initial product state of the form $|10\rangle$ as in Fig. 13 we merely span a two-dimensional space with basis states $\{|10\rangle, |01\rangle\}$. From this observation we can conjecture that it is the dimensionality of the Hilbert space of the auxiliary modes that determines our ability to design quantum networks with the least possible resources.

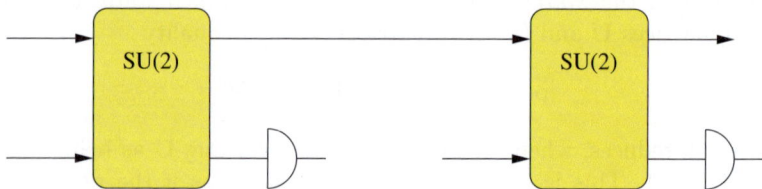

Fig. 13. Nonlinear sign shift gate with two beam splitters

3.4 Multi-Mode Gates

Similarly to the construction of single-mode quantum gates such as the nonlinear phase shift, we can proceed to more complex networks by referring to the results obtained before. The idea is rather simple. Let us restrict our atttention again to quantum operations within a certain Fock layer, i.e. a subspace of the total Hilbert space of a multimode system with fixed photon number. Then we construct an M-mode quantum gate in the following way:

1. first feed the M modes into a generalized Mach–Zehnder interferometer with M input and output ports ($2M$-port for short),
2. then act upon each output mode with an appropriate single-mode gate,
3. and finally recombine the M modes at another $2M$-port.

Note that the first and last steps are done deterministically.

Consider for example the controlled-phase gate whose truth table is given in Table 2. Because it represents a two-mode gate acting in the Fock layer with total photon number equal to two, this gate serves as the prime example for our statement. In terms of photonic amplitude operators, the conditional operator associated with the controlled-phase gate is given by

$$\hat{C}_\varphi = 1 - (1 - e^{i\varphi})\hat{n}_1\hat{n}_2 , \tag{31}$$

where we have already written the operator in the simplest form that matches the dimensionality of the signal-mode Hilbert spaces.[6] We can decompose the operator $\hat{C}_\varphi$ into a product $\hat{C}_\varphi = \hat{U}(\hat{N}_1 \otimes \hat{N}_2)\hat{U}^\dagger$ of two unitary operators $\hat{U}$ and $\hat{U}^\dagger$ associated with the Mach–Zehnder interferometer and a tensor product $\hat{N}_1 \otimes \hat{N}_2$ of two single-mode conditional operators corresponding to nonlinear phase shifts. Each of these has the form

$$\hat{N}_i = 1 - \frac{1}{2}(1 - e^{i\varphi})\hat{n}_i(\hat{n}_i - 1) , \quad i = 1, 2 . \tag{32}$$

Thus, combining all results we have obtained so far, we end up with the network depicted in Fig. 14. With the knowledge about the maximal success probability for a nonlinear sign flip (a nonlinear phase shift with $\varphi = \pi$) being $1/4$, the corresponding success probability of the controlled-phase gate will be $1/16$.

As we have already mentioned, this is not the highest probability one can actually achieve with linear optics and photodetection. The present "record" is set by [26] where a network has been found that works with a probability of $2/27$. This network does not have the structure we have discussed here.

3.5 Conditional Dynamics and Scaling of Success Probabilities

In connection with the original proposal for using post-selection of sub-dynamics it has been suggested that the use of conditional dynamics could eventually yield success probabilities that are arbitrarily close to unity. What

---

[6]Here we assume that the two-mode signal state is indeed only formed by the basis states $\{|00\rangle, |10\rangle, |01\rangle, |11\rangle\}$ without contributions from multi-photon states.

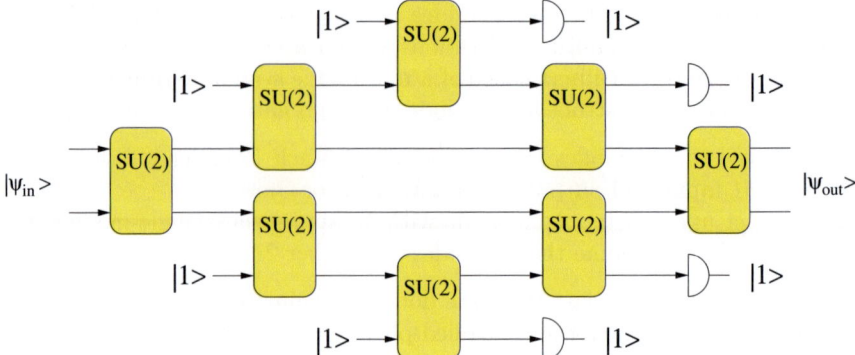

Fig. 14. Controlled-phase gate as two nonlinear phase shifts inside a balanced Mach–Zehnder interferometer

is meant by conditional dynamics is that depending on a certain measurement outcome (or pattern for more detectors) the resulting output state is either retained (in case the measurement gave the wanted answer) or post-processed using the knowledge about the transformation due to the detection result. We will give a simple argument that within the framework of qubits being encoded in photon numbers this cannot hold.

For this purpose, let us return to the single-mode gate described earlier. Suppose therefore we had a single-mode quantum state of the form $c_0|0\rangle + c_1|1\rangle + c_2|2\rangle$ at hand and we send it through the arrangement of beam splitters as depicted in Fig. 13. After the first beam splitter which is fed with an auxiliary single photon the output can contain as much as three photons in one output port. Hence, there is a certain probability that two or three photons are found in the photodetector. Let us look at the extreme case in which all three photons end up in the detector. By conservation of photon numbers, the output state contains only vacuum. But that means that we have *lost all information* about the signal state which was originally encoded in the weights c_i of the superposition. If we detected two photons instead we would lose the information contained in c_0. Thus, we can conclude that conditional dynamics breaks down if the total of number of photons measured in all photodetectors exceeds the total number of photons in the auxiliary state. Having said that, the obvious way of rescuing this situation is by noting that we have taken the measurements to be *projective* measurements. If in future one finds a way of performing non-destructive (QND) measurements, one has circumvented the information loss since all that has happened is a unitary transformation of the signal state.

For now, we have to live with projective measurements and it makes sense to ask what probabilities can be achieved in principle for certain gates and how they scale with increasing dimensionality of the signal state. We have already seen that the nonlinear sign flip can be realized with $p = 1/4$. In view of the algorithm to design multi-mode gates using generalized Mach–Zehnder interferometers, we could ask what are the respective success probabilities if we started from an $N + 1$-dimensional signal state of the form $c_0|0\rangle + c_1|1\rangle + \ldots + c_N|N\rangle$ and we would like to design a network that realizes a sign flip on the coefficient c_N. Then it turns out that if one used an N-dimensional auxiliary state with the lowest possible photon numbers, i.e. photon numbers in the range $0 \ldots N - 1$, there are networks that achieve a the desired transformation with a probability of $1/N^2$ [27].

4 Decoherence Mechanisms – QED in Causal Dielectric Media

"Please mind the gap between the hope and the reality."
— Smoke #2

4.1 Decoherence Mechanisms Affecting Atoms and Photons

Trapping Losses Near Current-Carrying Wires

As discussed in Sect. 2, the time the trapped atoms spend above the current-carrying wire is limited by several loss and heating mechanisms. In [13] one can find a table with several loss mechanisms relevant for atom chip experiments. We summarize the most important ones in Table 3 and give their approximate associated lifetimes. Most of the loss mechanisms are in fact avoidable as they are merely technical noise or account for experimental imperfections. Technical noise can cause both heating and spin flips. Heating processes are associated with low-frequency noise that causes the atoms in the trap to be excited into higher vibrational modes. Spin flips, on the other hand, are induced by rf noise. Suppose an atom stayed 1s over the wire. The Rabi frequency that induces the spin flips is therefore 0.5 Hz, which in turn is approximately given by 10^6 Hz/G times the magnetic field strength. Therefore, the rf magnetic field must be less than 1 μG which, for an atom being 1mm away from the wire, amounts to current fluctuations of less than 1 μA. For a current of 1–10A needed to produce the dc magnetic field this means that the power supply must be unusually stable. This is challenging but achievable.

Table 3. Loss mechanisms for trapped atoms above current-carrying wires

Mechanism	Lifetime
thermally induced spin flips	$1 - 10$ s
technical noise	> 10 s
background collisions	> 10 s

However, some loss mechanisms are inherent in the way the atoms are trapped and can hardly be avoided. The most notable of these unavoidable losses is that caused by thermally induced spin flips which incidentally already gives rise to the biggest loss rate in current experiments [28]. In Fig. 15 we show the five magnetic sublevels of the $|F = 2\rangle$ state of ^{87}Rb. In the experiment, the atoms are optically pumped into the $|F=2, m=2\rangle$ sublevel in which they are trapped. Spin flips induced by fluctutating magnetic fields cause the atom to jump into lower-lying magnetic sublevels in which they are either lost due to gravitational forces ($|F = 2, m = 0\rangle$) or expelled from the trap ($|F = 2, m = -1\rangle$ and $|F = 2, m = -2\rangle$). These spin-flip transitions are caused by magnetic-field fluctuations that are unavoidable as soon as metallic or dielectric materials are located close to the atoms. This is quite obviously the case in atom chips. In order to see where these fluctuations come from, we have to look deeper into the statistical implications of macroscopic (quantum) electrodynamics.

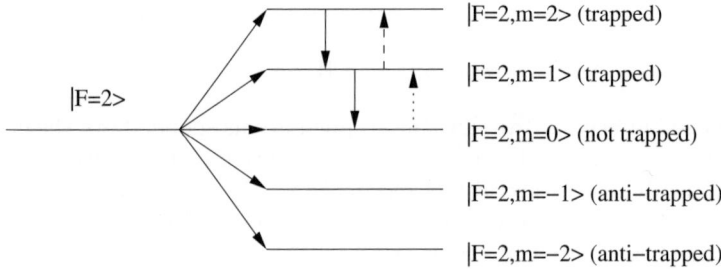

Fig. 15. Zeeman splitting of the $|F = 2\rangle$ state of ^{87}Rb and some of the possible spin-flip transitions

Gate Fidelity with Imperfect Linear Optical Elements

In Sect. 3, which deals with all-optical realisations of quantum information processing with passive linear optical elements, we have seen that the fidelity of a gate operation will depend crucially on the ability to generate single photons with very high efficiency and the amount of losses that are introduced by imperfect passive elements and photodetectors. It is not difficult to imagine that the purity of the ancilla photons used to operate a quantum circuit will have enormous influence on the way the quantum gates operate. For example, an ancilla state consisting of a *mixture* of vacuum and a single-photon Fock state actually corresponds to realising *two different* quantum gates at once or, in other words, a statistical mixture of their respective operations. Another example of a decoherence process is provided by the inherent absorption in the passive linear optical elements used to build up the quantum circuit. An absorbing beam splitter does *not* act unitarily on the electromagnetic field modes impinging on it, but rather as a general CP map, i.e. as a statistical mixture of the desired and some unwanted operations.

All the loss and decoherence mechanisms sketched above can be described by a single and powerful theory which we will outline in this section.

4.2 Field Quantisation in Causal Media

"Die ganzen Jahre bewußter Grübelei haben mich der Antwort der Frage 'Was sind Lichtquanten' nicht näher gebracht. Heute glaubt zwar jeder Lump, er wisse es, aber er täuscht sich."
— Albert Einstein, in a letter to Michele Besso (1951)

Quantum electrodynamics (or QED for short) is a well-established theory, and one might wonder how it is possible to obtain fundamentally new results apart from those that are already known. The answer comes about when considering *phenomenological* electrodynamics and when trying to quantize it.

Classical Electrodynamics with Media

Maxwell's equations of the electromagnetic field in free space, i.e. without external sources, in temporal Fourier space can be cast in the following form:

$$\boldsymbol{\nabla} \cdot \mathbf{B}(\mathbf{r}, \omega) = 0 \,,$$
$$\boldsymbol{\nabla} \cdot \mathbf{D}(\mathbf{r}, \omega) = 0 \,,$$
$$\boldsymbol{\nabla} \times \mathbf{E}(\mathbf{r}, \omega) = i\omega \mathbf{B}(\mathbf{r}, \omega) \,,$$
$$\boldsymbol{\nabla} \times \mathbf{H}(\mathbf{r}, \omega) = -i\omega \mathbf{D}(\mathbf{r}, \omega) \,. \tag{33}$$

These equations have to be supplemented by appropriate constitutive relations which relate the electric field $\mathbf{E}(\mathbf{r}, \omega)$ and the magnetic induction $\mathbf{B}(\mathbf{r}, \omega)$ to the displacement field $\mathbf{D}(\mathbf{r}, \omega)$ and the magnetic field $\mathbf{H}(\mathbf{r}, \omega)$ which also carry information about the material. Assuming a purely dielectric medium, i.e. disregarding magnetic properties, one usually defines a polarisation field $\mathbf{P}(\mathbf{r}, \omega)$ via

$$\mathbf{D}(\mathbf{r}, \omega) = \varepsilon_0 \mathbf{E}(\mathbf{r}, \omega) + \mathbf{P}(\mathbf{r}, \omega) \,. \tag{34}$$

In real space, and assuming locally and linearly responding media, the polarisation can be written as a temporal convolution

$$\mathbf{P}(\mathbf{r}, t) = \varepsilon_0 \int\limits_0^\infty d\tau \, \chi(\mathbf{r}, \tau) \mathbf{E}(\mathbf{r}, t - \tau) + \mathbf{P}_N(\mathbf{r}, t) \tag{35}$$

with the linear susceptibility $\chi(\mathbf{r}, \tau)$, the Fourier transform of which is related to the dielectric permittivity $\varepsilon(\mathbf{r}, \omega)$ by

$$\varepsilon(\mathbf{r}, \omega) = \chi(\mathbf{r}, \omega) + 1 \,. \tag{36}$$

The permittivity is a complex function of frequency, $\varepsilon(\mathbf{r}, \omega) = \varepsilon_R(\mathbf{r}, \omega) + i\varepsilon_I(\mathbf{r}, \omega)$. The real and imaginary parts, which are responsible for dispersion and absorption, respectively, are related to each other by the Kramers–Kronig relations

$$\varepsilon_R(\mathbf{r}, \omega) - 1 = \frac{1}{\pi} \mathcal{P} \int d\omega' \frac{\varepsilon_I(\mathbf{r}, \omega')}{\omega' - \omega}, \tag{37}$$

$$\varepsilon_I(\mathbf{r}, \omega) = -\frac{1}{\pi} \mathcal{P} \int d\omega' \frac{\varepsilon_R(\mathbf{r}, \omega') - 1}{\omega' - \omega} \tag{38}$$

with $\mathcal{P}$ denoting the principal value. The complex permittivity is an analytic function in the upper complex half-plane without zeros and satifies the relation

$$\varepsilon(\mathbf{r}, -\omega^*) = \varepsilon^*(\mathbf{r}, \omega). \tag{39}$$

Furthermore, it approaches unity in the high-frequency limit. These analyticity properties, which carry over to the Green function to be defined later, will be important for proving equal-time commutation relations (ETCR for short) between electromagnetic field operators. This list of properties can be shown to follow from causality.

Equation (35) contains an additional, non-convolutive term $\mathbf{P}_N(\mathbf{r}, t)$. In fact, the existence of this term is crucial for the functioning of the following quantisation procedure. It serves as a Langevin noise source which is needed to preserve the ETCR. From statistical physics it is known that dissipation processes (as described by the imaginary part of the dielectric permittivity) are always accompanied by additional fluctuations. In order to see this more clearly, assume a damped harmonic oscillator with the solution for the expectation values $\langle \hat{a}(t) \rangle = \langle \hat{a}(t') \rangle e^{-\Gamma(t-t')}$. This relation cannot be valid in operator form since with increasing time the ETCR between $\hat{a}(t)$ and $\hat{a}^\dagger(t)$ would decay exponentially and eventually violates Heisenberg's uncertainty relations. It is possible to add a Langevin force with vanishing expectation value to the oscillator which takes care of the ETCR. The noise polarisation $\mathbf{P}_N(\mathbf{r}, t)$ is exactly such a Langevin force. Its strength is determined by the linear fluctuation-dissipation theorem [29]. Assuming that $\mathbf{P}_N(\mathbf{r}, t)$ is proportional to a Gaussian random variable $\mathbf{f}(\mathbf{r}, \omega)$, which is always possible in linear reponse theory, and noting that the linear fluctuation-dissipation theorem states that the correlation function $\langle \mathbf{P}(\mathbf{r}, \omega)\mathbf{P}(\mathbf{r}', \omega') \rangle$ is proportional to the imaginary part of the response function – hence the dielectric permittivity – we immediately deduce that the correct form of the noise polarisation must be [20, 30, 31, 32]

$$\mathbf{P}_N(\mathbf{r}, t) = i\sqrt{\frac{\hbar\varepsilon_0}{\pi} \varepsilon_I(\mathbf{r}, \omega)}\, \mathbf{f}(\mathbf{r}, \omega). \tag{40}$$

As a matter of fact, the $\mathbf{f}(\mathbf{r}, \omega)$ plays the rôle of the fundamental variable in terms of which all relevant quantities can be expressed. Consider for example the Helmholtz equation for the electric field which is obtained by substituting Faraday's law (33) into Amperè's law (33) and using the constitutive relation (34) as

$$\nabla \times \nabla \times \mathbf{E}(\mathbf{r}, \omega) - \frac{\omega^2}{c^2} \varepsilon(\mathbf{r}, \omega) \mathbf{E}(\mathbf{r}, \omega) = \mu_0 \omega^2 \mathbf{P}_N(\mathbf{r}, \omega) \,. \tag{41}$$

The solution to (41) is easily found to be

$$\mathbf{E}(\mathbf{r}, \omega) = \mu_0 \omega^2 \int d^3 s \, \mathbf{G}(\mathbf{r}, \mathbf{s}, \omega) \cdot \mathbf{P}_N(\mathbf{s}, \omega) \tag{42}$$

where the (tensor-valued) Green function satifies

$$\nabla \times \nabla \times \mathbf{G}(\mathbf{r}, \mathbf{s}, \omega) - \frac{\omega^2}{c^2} \varepsilon(\mathbf{r}, \omega) \mathbf{G}(\mathbf{r}, \mathbf{s}, \omega) = \delta(\mathbf{r} - \mathbf{s}) \mathbf{U} \,, \tag{43}$$

where $\mathbf{U}$ denotes the unit dyad. Equation (42) shows explictly how all field quantities can eventually be expressed, via the noise polarisation, in terms of the fundamental field $\mathbf{f}(\mathbf{r}, \omega)$. As mentioned earlier, the dyadic Green function $\mathbf{G}(\mathbf{r}, \mathbf{s}, \omega)$, being a causal response function itself, has all the analytic properties the dielectric permittivity has, i.e. it is holomorphic in the upper complex frequency half-plane and satifies the relation

$$\mathbf{G}(\mathbf{r}, \mathbf{s}, -\omega^*) = \mathbf{G}^*(\mathbf{r}, \mathbf{s}, \omega) \,. \tag{44}$$

Moreover, it fulfils a generalized Onsager–Machlup reciprocity [29] relation of the form

$$\mathbf{G}(\mathbf{s}, \mathbf{r}, \omega) = \mathbf{G}^T(\mathbf{r}, \mathbf{s}, \omega) \,. \tag{45}$$

From the partial differential equation (43) it also follows that the dyadic Green function satisfies an important integral relation,

$$\int d^3 s \, \frac{\omega^2}{c^2} \varepsilon_I(\mathbf{s}, \omega) \mathbf{G}(\mathbf{r}, \mathbf{s}, \omega) \cdot \mathbf{G}^+(\mathbf{r}', \mathbf{s}, \omega) = \mathrm{Im} \mathbf{G}(\mathbf{r}, \mathbf{r}', \omega) \,, \tag{46}$$

which expresses the fluctuation-dissipation theorem.

Quantized Electrodynamics with Media

The theory described above can now be quantized by regarding the fundamental field $\mathbf{f}(\mathbf{r}, \omega)$ as a bosonic vector field $\hat{\mathbf{f}}(\mathbf{r}, \omega)$ with the commutation relation

$$\left[\hat{\mathbf{f}}(\mathbf{r}, \omega), \hat{\mathbf{f}}^\dagger(\mathbf{r}', \omega') \right] = \delta(\mathbf{r} - \mathbf{r}') \delta(\omega - \omega') \mathbf{U} \,. \tag{47}$$

Then, for example, the Fourier component of the electric field operator can be written, on using (42), as

$$\hat{\mathbf{E}}(\mathbf{r}, \omega) = i \sqrt{\frac{\hbar}{\pi \varepsilon_0} \frac{\omega^2}{c^2}} \int d^3 s \, \sqrt{\varepsilon_I(\mathbf{s}, \omega)} \mathbf{G}(\mathbf{r}, \mathbf{s}, \omega) \cdot \hat{\mathbf{f}}(\mathbf{s}, \omega) \,. \tag{48}$$

The corresponding Schrödinger operator is obtained by integrating over all frequencies,

$$\hat{\mathbf{E}}(\mathbf{r}) = \int\limits_0^\infty d\omega\, \hat{\mathbf{E}}(\mathbf{r},\omega) + \text{h.c.} . \tag{49}$$

The so-defined operators fulfil all the requirements one imposes on a statistical quantum field theory. Firstly, the ETCR between electric field and magnetic induction can be shown to be [30, 31, 32]

$$\left[\hat{\mathbf{E}}(\mathbf{r}), \hat{\mathbf{B}}(\mathbf{r}')\right] = -\frac{i\hbar}{\varepsilon_0}\boldsymbol{\nabla} \times \delta(\mathbf{r} - \mathbf{r}')\boldsymbol{U} . \tag{50}$$

Secondly, the theory is – by construction – also consistent with the fluctuation-dissipation theorem which we can cast in the form

$$\langle 0|\hat{\mathbf{E}}(\mathbf{r},\omega)\hat{\mathbf{E}}^\dagger(\mathbf{r}',\omega')|0\rangle = \frac{\hbar\omega^2}{\pi\varepsilon_0 c^2}\text{Im}\boldsymbol{G}(\mathbf{r},\mathbf{r}',\omega)\delta(\omega - \omega') . \tag{51}$$

Equation (51) tells us that the (vacuum) fluctuations of the electric field are given by the imaginary part of the Green function as one would expect. Finally, the Hamiltonian describing the electromagnetic field in the presence of absorbing matter is given by

$$\hat{H} = \int d^3\mathbf{r} \int\limits_0^\infty d\omega\, \hbar\omega\, \hat{\mathbf{f}}^\dagger(\mathbf{r},\omega) \cdot \hat{\mathbf{f}}(\mathbf{r},\omega) \tag{52}$$

which is diagonal in the fundamental bosonic vector field $\hat{\mathbf{f}}(\mathbf{r},\omega)$. Maxwell's equations then follow from the Heisenberg equations of motion of the above-defined electromagnetic field operators.

Extensions of this theory to spatially anisotropic materials (for which the dielectric permittivity is a symmetric tensor) and media with gain can be found in [31, 32]. Magnetic materials, including the important class of magnetodielectrics or left-handed media, can be treated in a completely analogous way [33]. We will, however, not elaborate more on this subject as it is not needed for our present purposes.

4.3 Thermally Induced Spin Flips Near Metallic Wires

The above-described theory is perfectly suited for describing magnetic-field fluctuations near metallic surfaces, which cause spin flips between trapped and anti-trapped (or non-trapped) magnetic sublevels of atomic hyperfine ground states. The experiment in [28] used a coated wire to trap ^{87}Rb atoms in their $|F=2, m=2\rangle$ magnetic sublevel. The atoms are located at a distance r from the surface of the wire with radius a_2 (see Fig. 16).

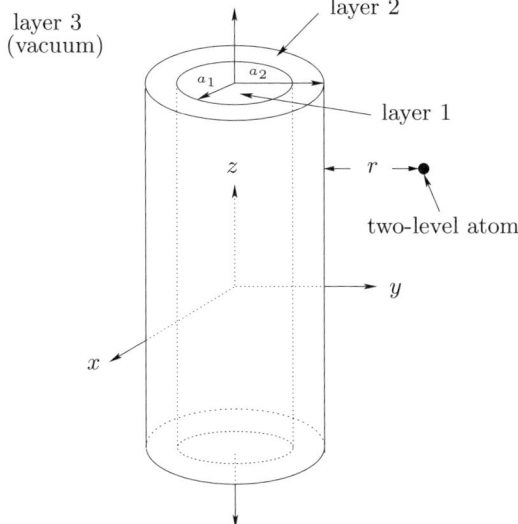

Fig. 16. A two-level atom is located at a distance r from the surface of a coated wire that runs along the z-direction

The interaction of a neutral atom positioned at some point $\mathbf{r}_A$ with the electromagnetic field described by the Hamiltonian (52) via the Zeeman interaction $\hat{H}_Z = -\hat{\boldsymbol{\mu}} \cdot \hat{\mathbf{B}}(\mathbf{r}_A)$ of the atom's magnetic moment $\hat{\boldsymbol{\mu}}$ with the magnetic field causes the spin of the atoms to flip. Since the alkali atoms are in their electronic ground state, the orbital angular momentum is zero. Moreover, the nuclear magnetic moment is small compared to the Bohr magneton μ_B. This leaves us with an expression for the atom's magnetic moment associated with the transition $|i\rangle \rightarrow |f\rangle$ as $\hat{\boldsymbol{\mu}} = \boldsymbol{\mu}|i\rangle\langle f| + \text{h.c.}$ where the magnitude $\boldsymbol{\mu}$ is given by the (electronic) spin matrix element $\boldsymbol{\mu} = 2\mu_B\langle i|\hat{\mathbf{S}}|f\rangle$. If we define the spin flip operators $\hat{\sigma} = |f\rangle\langle i|$ satisfying the angular-momentum algebra $[\hat{\sigma}, \hat{\sigma}_z] = \hat{\sigma}$, we can write the total Hamiltonian in rotating-wave approximation as

$$
\hat{H} = \int d^3\mathbf{r} \int_0^\infty d\omega\, \hbar\omega\, \hat{\mathbf{f}}^\dagger(\mathbf{r}, \omega) \cdot \hat{\mathbf{f}}(\mathbf{r}, \omega) + \frac{1}{2}\hbar\omega_{fi}\hat{\sigma}_z
$$

$$
- \left[\hat{\sigma}^\dagger \boldsymbol{\mu} \cdot \boldsymbol{\nabla} \times \int_0^\infty d\omega \int d^3\mathbf{r}\, \frac{\omega}{c^2} \sqrt{\frac{\hbar}{\pi\varepsilon_0}} \varepsilon_I(\mathbf{r}, \omega) \mathbf{G}(\mathbf{r}_A, \mathbf{r}, \omega) \cdot \hat{\mathbf{f}}(\mathbf{r}, \omega) + \text{h.c.} \right]
$$
(53)

Using the Hamiltonian (53) in the Heisenberg equations of motion for the spin flip operators, we find that, in the Markov approximation, the spin flip rate is given by [34]

$$\Gamma = \frac{2g_S^2\mu_B^2}{\hbar\varepsilon_0 c^2} \langle f|\hat{\mathbf{S}}|i\rangle \cdot \text{Im}\left[\boldsymbol{\nabla} \times \boldsymbol{\nabla} \times \boldsymbol{G}(\mathbf{r}_A,\mathbf{r}_A,\omega_{fi})\right] \cdot \langle i|\hat{\mathbf{S}}|f\rangle . \qquad (54)$$

Taking into account that the flip rate is thermally enhanced, i.e. not only spontaneous but also stimulated (induced) flips occur, the rate (54) has to be modified by a multiplicative factor $n_{th} + 1$, where $n_{th} = (e^{\hbar\omega_{fi}/kT} - 1)^{-1}$ which is proportional to T for high temperature. Furthermore, considering the initial state $|i\rangle$ to be the $|F=2, m=2\rangle$ magnetic sublevel and the final state $|f\rangle$ to be the state $|F=2, m=1\rangle$, the transition matrix element squared gives a factor of $1/8$ due to the expansion of the $|F, m\rangle$-states in terms of the eigenstates of electronic and nuclear spin operators. We are now in a position to compare with experimental measurements. The experiment in [28] used a coated wire at a temperature $T = 380\,\text{K}$ with a copper core of $185\,\mu\text{m}$ radius and an aluminium coating with $55\,\mu\text{m}$ thickness. The transition frequency between two neighboring magnetic sublevels was $f = \omega_{fi}/2\pi = 560\,\text{kHz}$. The result of the calculations with our theory and the experimental data are shown in Fig. 17 where we have plotted the inverse spin flip rate, hence the average lifetime τ of an atom in the trap. It is clear that our theoretical result will only serve as an upper bound on the lifetime since, as we have mentioned earlier, there are indeed other noise sources that influence the duration of stay of the atoms in their trap. Note, however, that the agreement is still astonishingly good which means that our earlier claim that most other noise source are negligible in this experiment, is indeed valid.

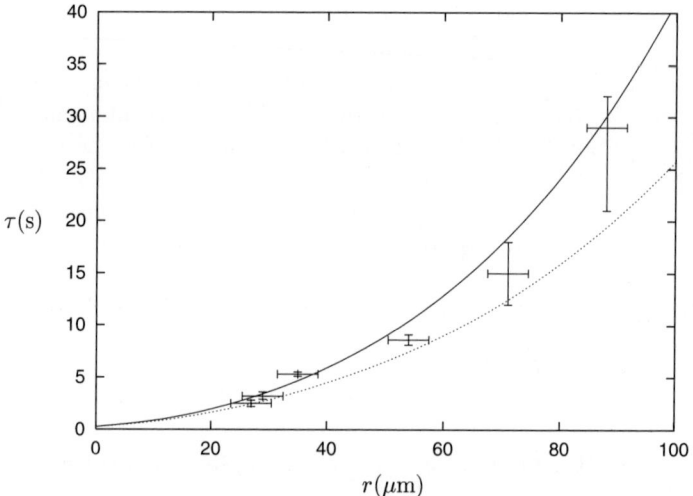

Fig. 17. Lifetime τ of the trapped atom as a function of the distance r from the surface of the current-carrying wire. The *solid line* corresponds to the calculated lifetime according to the theory presented in the text. The crosses represent experimental data points. The *dotted line* is the result for a slab-geometry according to [35]

4.4 Imperfect Passive Optical Elements

Another, seemingly unrelated, application of the field quantization in absorbing materials is found if we consider passive optical elements such as beam splitters under realistic conditions, i.e. including losses in the beam splitter material. As noted earlier, a lossless beam splitter acts upon the amplitude operators of the incoming waves as an SU(2) group element [18]. That is, two amplitude operators $\hat{a}_1(\omega)$ and $\hat{a}_2(\omega)$ are transformed into new amplitude operators $\hat{b}_1(\omega)$ and $\hat{b}_2(\omega)$ as

$$
\begin{pmatrix} \hat{b}_1(\omega) \\ \hat{b}_2(\omega) \end{pmatrix} = \mathbf{T}(\omega) \begin{pmatrix} \hat{a}_1(\omega) \\ \hat{a}_2(\omega) \end{pmatrix}
\tag{55}
$$

where $\mathbf{T}(\omega)$ is a unitary matrix, i.e. $\mathbf{T}(\omega)\mathbf{T}^+(\omega) = \mathbf{I}$. For lossy beam splitters, this is certainly not true since some of the electromagnetic radiation impinging on the beam splitter will actually be absorbed. This process can be accounted for by introducing an absorption matrix $\mathbf{A}(\omega)$ in order to satisfy the energy conservation $\mathbf{T}(\omega)\mathbf{T}^+(\omega) + \mathbf{A}(\omega)\mathbf{A}^+(\omega) = \mathbf{I}$. Both matrices actually follow from a decomposition of the associated Green function into contributions associated with travelling waves moving to the left and right [36] and are therefore given by the complex refractive-index profile of the beam splitter material and geometric parameters.

However, it turns out that, associating the material excitations with some bosonic variables $\hat{g}_1(\omega)$ and $\hat{g}_2(\omega)$ which are special linear combinations of the fundamental bosonic field variables $\hat{\mathbf{f}}(\mathbf{r}, \omega)$ restricted to one spatial dimension [36], the beam splitter acts unitarily on the combined set of photonic and material amplitude operators [37]. In fact, a lossy beam splitter can be represented as an SU(4) group element. That is, the 'four-vector' $\hat{\boldsymbol{\alpha}}(\omega)$ consisting of incoming photonic amplitude operators and the material excitations transforms unitarily into the 'four-vector' $\hat{\boldsymbol{\beta}}(\omega)$ of outgoing photonic amplitude operators and some new matter operators, i.e.

$$
\hat{\boldsymbol{\beta}}(\omega) = \boldsymbol{\Lambda}(\omega)\hat{\boldsymbol{\alpha}}(\omega), \quad \boldsymbol{\Lambda}(\omega)\boldsymbol{\Lambda}^+(\omega) = \boldsymbol{I} .
\tag{56}
$$

The unitary 4×4-matrix $\boldsymbol{\Lambda}(\omega)$ can be decomposed into block form as

$$
\boldsymbol{\Lambda}(\omega) = \begin{pmatrix} \mathbf{T}(\omega) & \mathbf{A}(\omega) \\ -\mathbf{S}(\omega)\mathbf{C}^{-1}(\omega)\mathbf{T}(\omega) & \mathbf{C}(\omega)\mathbf{S}^{-1}(\omega)\mathbf{T}(\omega) \end{pmatrix}
\tag{57}
$$

where $\mathbf{C}(\omega) = \sqrt{\mathbf{T}(\omega)\mathbf{T}^+(\omega)}$ and $\mathbf{S}(\omega) = \sqrt{\mathbf{A}(\omega)\mathbf{A}^+(\omega)}$ are commuting positive Hermitian 2×2-matrices. The associated unitary operator $\hat{U}$ that realizes the transformation $\hat{\boldsymbol{\beta}}(\omega) = \boldsymbol{\Lambda}(\omega)\hat{\boldsymbol{\alpha}}(\omega) = \hat{U}^\dagger \hat{\boldsymbol{\alpha}}(\omega)\hat{U}$ is then obtained as

$$
\hat{U} = \exp\left\{ -i \int_0^\infty d\omega \, [\hat{\boldsymbol{\alpha}}^\dagger(\omega)]^T \, \boldsymbol{\Phi}(\omega)\hat{\boldsymbol{\alpha}}(\omega) \right\}
\tag{58}
$$

where the matrix $\boldsymbol{\Phi}(\omega)$ is defined by $e^{-i\boldsymbol{\Phi}(\omega)} = \boldsymbol{\Lambda}(\omega)$. The operator $\hat{U}$ and the matrix $\boldsymbol{\Lambda}(\omega)$ can be used as well to transform quantum states. For this purpose, let us assume that the density operator of the quantum state before the transformation is an operator functional of $\hat{\boldsymbol{\alpha}}(\omega)$, i.e. $\hat{\varrho} = \hat{\varrho}[\hat{\boldsymbol{\alpha}}(\omega), \hat{\boldsymbol{\alpha}}^{\dagger}(\omega)]$. Then, transformation of the quantum state and tracing over the degrees of freedom associated with the beam splitter leaves us with

$$\hat{\varrho}' = \mathrm{Tr}\left\{\hat{\varrho}\left[\hat{U}\hat{\boldsymbol{\alpha}}(\omega)\hat{U}^{\dagger}, \hat{U}\hat{\boldsymbol{\alpha}}^{\dagger}(\omega)\hat{U}^{\dagger}\right]\right\}. \tag{59}$$

In quantum information theory, decoherence processes are frequently described by the Kraus decomposition of the quantum-state transformation (59). That is, we seek an operator decomposition of the form

$$\hat{\varrho}' = \sum_{i} \hat{W}_{i}\hat{\varrho}\hat{W}_{i}^{\dagger}, \quad \sum_{i} \hat{W}_{i}^{\dagger}\hat{W}_{i} = \hat{I}. \tag{60}$$

For this purpose, note that the beam splitter is in its ground state, thereby neglecting thermal excitations which at room temperature and optical frequencies is a safe assumption. Let us also restrict ourselves to quasi-monochromatic radiation in which case we can drop all frequency dependencies. Then, calculating the partial trace in (59) in the coherent-state basis as

$$\hat{\varrho}' = \frac{1}{\pi^2} \int d^2\alpha_3 d^2\alpha_4 \, \hat{W}_{\alpha_3,\alpha_4} \hat{\varrho} \hat{W}_{\alpha_3,\alpha_4}^{\dagger} \tag{61}$$

where the continuous-index Kraus operators are defined by

$$\hat{W}_{\alpha_3,\alpha_4} = \langle \alpha_3, \alpha_4 | \hat{U} | 0_3, 0_4 \rangle, \tag{62}$$

we eventually find that the Kraus operators of an absorbing beam splitter are $(\exp{-i\boldsymbol{\Phi}_T} = \mathbf{T})$ [23]

$$\hat{W}_{\alpha_3,\alpha_4} = \exp\left(-i[\hat{\mathbf{a}}^{\dagger}]^T \boldsymbol{\Phi}_T \hat{\mathbf{a}}\right) \exp\left(-\boldsymbol{\alpha}^{+}\mathbf{SC}^{-1}\mathbf{T}\hat{\mathbf{a}}\right) \exp\left(-\boldsymbol{\alpha}^{+}\boldsymbol{\alpha}/2\right). \tag{63}$$

One checks easily that these operators become unitary operators in the limit of vanishing absorption, i.e. when $\mathbf{T}$ is unitary and therefore $\boldsymbol{\Phi}_T$ Hermitian, and $\mathbf{S}$ vanishes. We also see that the term $\exp(-\boldsymbol{\alpha}^{+}\mathbf{SC}^{-1}\mathbf{T}\hat{\mathbf{a}})$ is responsible for absorption as it is a function of the annihilation operators only.

Other Error Sources

In addition to absorption in the dielectric material, there are several other sources of errors in an experimental implementation. So far, we have assumed that the auxiliary states are perfectly pure states, especially we have implicitly assumed that we are able to produce single photons on demand with very high efficiency. This, however, is an oversimplification. All realistic sources produce at most a mixed state of the form $(1-p)|0\rangle\langle 0| + p|1\rangle\langle 1|$ with efficiency

p. The best single-photon sources to date achieve a single-photon efficiency of at most 80%. This means that the first beam splitter in Fig. 13 does not always act as a first-order polynomial in $\hat{n}$, but sometimes subtracts a photon from the signal state.

The detectors themselves are a crucial error source. In our mathematical description we have assumed that the photodetectors are perfect and can therefore be described by a projection operator $|n\rangle\langle n|$. However, since they show losses themselves, we have to replace it by

$$|n\rangle\langle n| \mapsto \hat{\Pi}(n) = \sum_{k=n}^{\infty} \binom{k}{n} \eta^n (1-\eta)^{k-n} |k\rangle\langle k| \tag{64}$$

which constitutes a positive operator-valued measure (a generalization of a projective measurement). It describes the effect of absorption in the photodetector itself. The number η is commonly called the detector efficiency. Typical values for avalanche photo diodes are $\eta \approx 0.3$.

We can measure the effect of these error sources by introducing the gate fidelity F as the overlap between the wanted result $|\psi'\rangle$ and the achieved density matrix $\hat{\varrho}$, i.e. $F = \langle\psi'|\hat{\varrho}|\psi'\rangle$. Since the so-defined fidelity still depends on the chosen signal state – because not all signal states are effected in the same manner by the errors – we define an average fidelity $\bar{F}$ by averaging over all possible signal states.

Figure 18 shows the effect of non-unit single-photon efficiency p (left figure) and detector efficiency η (right figure) on the average gate fidelity. One can see that even a small inefficiency in either the detection process or the preparation process of the auxiliary state results in a (roughly linear) drop in the gate fidelity. In a different setting, in which the quantum information is encoded in the superposition states of one photonic excitation in two modes, a similar behaviour has been observed [38]. This result has important consequences for the scalability of such gates since quantum error correction only works if the individual errors in the network components does not exceed $\approx 10^{-3}$ (see for example [39] and references cited therein). In other words, in

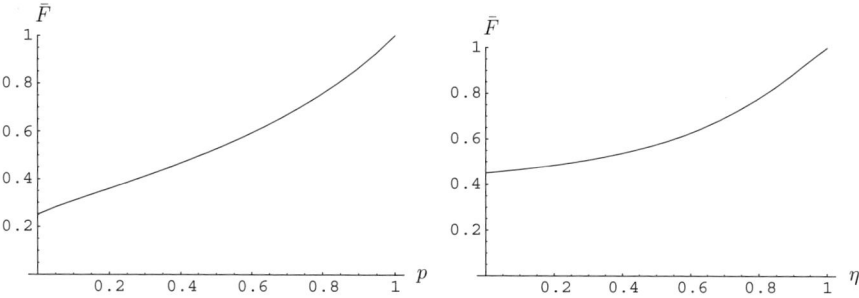

Fig. 18. Average gate fidelity with imperfect auxiliary photons (*left figure*) and imperfect detectors (*right figure*)

order to be able to correct for errors, both discussed inefficiencies must be at most of the same order as the desired error rate which currently seems to be experimentally impossible.

Acknowledgements

We would like to thank Dominic Berry, Matthew Jones, Norbert Lütkenhaus, Bill Munro, Kae Nemoto, Per–Kristian Rekdal and Barry Sanders for discussions and their input to the research reviewed here. This work was funded in parts by the UK Engineering and Physical Sciences Research Council (EPSRC), the Royal Society, and the European Commission (QUEST, QGATES, and FASTNET networks). Special thanks goes to John, publican of the Gowlett Arms, for providing one us with the necessary amount of Brains.

References

1. H. Rabitz, R. de Vivie–Riedle, M. Motzkus, K. Kompa: Science **288**, 824 (2000); C.H. Bennett, J.I. Cirac, M.S. Leifer, D.W. Leung, N. Linden, S. Popescu, G. Vidal: Phys. Rev. A **66**, 012305 (2002)
2. V. Bužek, M. Hillery, R.F. Werner: Phys. Rev. A **60**, R2626 (1999)
3. D.P. DiVincenzo: Phys. Rev. A **51**, 1015 (1995)
4. D. Jaksch, C. Bruder, J.I. Cirac, C.W. Gardiner, P. Zoller: Phys. Rev. Lett. **81**, 3108 (1998)
5. M. Greiner, I. Bloch, O. Mandel, T.W. Hänsch, T. Esslinger: Phys. Rev. Lett. **87**, 160405 (2001); M. Greiner, O. Mandel, T. Esslinger, T.W. Hänsch, I. Bloch: Nature (London) **415**, 39 (2002)
6. C. Orzel, A.K. Tuchman, M.L. Fenselau, M. Yasuda, M.A. Kasevich: Science **291**, 2386 (2001)
7. D.S. Petrov, G.V. Shlyapnikov, J.T.M. Walraven: Phys. Rev. Lett. **85**, 3745 (2000)
8. V. Dunjko, V. Lorent, M. Olshanii: Phys. Rev. Lett. **86**, 5413 (2001)
9. M.P.A. Fisher, P.B. Weichman, G. Grinstein, D.S. Fisher: Phys. Rev. B **40**, 546 (1989)
10. J. Pachos, P.L. Knight: Phys. Rev. Lett. **91**, 107902 (2003)
11. O. Mandel, M. Greiner, A. Widera, T. Rom, T.W. Hänsch, I. Bloch: Phys. Rev. Lett. **91**, 010407 (2003)
12. E.A. Hinds, I.A. Hughes: J. Phys. D: Appl. Phys. **32**, R119 (1999)
13. R. Folman, P. Krüger, J. Schmiedmayer, J. Denschlag, C. Henkel: Adv. At. Mol. Opt. Phys. **48**, 263 (2002)
14. E.A. Hinds: Phil. Trans. R. Soc. Lond. A **357**, 1409 (1999)
15. M. Koashi, T. Yamamoto, N. Imoto: Phys. Rev. A **63**, 030301 (2001)
16. T.B. Pittman, B.C. Jacobs, J.D. Franson: Phys. Rev. A **64**, 062311 (2001); T.B. Pittman, B.C. Jacobs, J.D. Franson: Phys. Rev. Lett. **88**, 257902 (2002)
17. E. Knill, R. Laflamme, G.J. Milburn: Nature (London) **409**, 46 (2001)

18. B. Yurke, S.L. McCall, J.R. Klauder: Phys. Rev. A **33**, 4033 (1986); S. Prasad, M.O. Scully, W. Martienssen: Opt. Commun. **62**, 139 (1987); Z.Y. Ou, C.K. Hong, L. Mandel: Opt. Commun. **63**, 118 (1987); H. Fearn, R. Loudon, Opt. Commun. **64**, 485 (1987); M.A. Campos, B.E.A. Saleh, M.C. Teich: Phys. Rev. A **40**, 1371 (1989); U. Leonhardt: Phys. Rev. A **48**, 3265 (1993)
19. K. Wodkiewicz, J.H. Eberly: J. Opt. Soc. Am. B **2**, 458 (1985)
20. W. Vogel, D.-G. Welsch, S. Wallentowitz: *Quantum Optics: An Introduction* (Wiley-VCH, Berlin, 2001)
21. M. Reck, A. Zeilinger, H.J. Bernstein, P. Bertani: Phys. Rev. Lett. **73**, 58 (1994)
22. M. Byrd: Preprint arXiv:physics/9708015; T. Tilma, M. Byrd, E.C.G. Sudarshan: J. Phys. A: Math. Gen. **35**, 10445 (2002)
23. S. Scheel, K. Nemoto, W.J. Munro, Peter L. Knight: Phys. Rev. A **68**, 032310 (2003)
24. H. Minc: *Permanents* (Addison-Wesley, London, 1978)
25. T.C. Ralph, A.G. White, W.J. Munro, G.J. Milburn: Phys. Rev. A **65**, 012314 (2001)
26. E. Knill: Phys. Rev. A **66**, 052306 (2002)
27. S. Scheel, N. Lütkenhaus: New J. Phys. **6**, 51 (2004); S. Scheel, K.M.R. Audenaert: New J. Phys. **7**, 149 (2005)
28. M.P.A. Jones, C.J. Vale, D. Sahagun, B.V. Hall, E.A. Hinds: Phys. Rev. Lett. **91**, 080401 (2003)
29. R.L. Stratonovich: *Nonlinear nonequilibrium thermodynamics I: Linear and nonlinear fluctuation-dissipation theorems* (Springer, Heidelberg, 1992)
30. T.D. Ho, L. Knöll, D.-G. Welsch: Phys. Rev. A **57**, 3931 (1998)
31. S. Scheel, L. Knöll, D.-G. Welsch: Phys. Rev. A **58**, 700 (1998)
32. L. Knöll, S. Scheel, D.-G. Welsch: QED in dispersing and absorbing dielectric media. In: *Coherence and Statistics of Photons and Atoms*, ed. by J. Peřina (Wiley, New York, 2001) pp. 1–64
33. T.D. Ho, S.Y. Buhmann, L. Knöll, D.G. Welsch, S. Scheel, J. Kästel: Phys. Rev. A **68**, 043816 (2003)
34. P.K. Rekdal, S. Scheel, E.A. Hinds, P.L. Knight: Phys. Rev. A **70**, 013811 (2004)
35. C. Henkel, S. Pötting, M. Wilkens: Appl. Phys. B **69**, 379 (1999)
36. T. Gruner, D.G. Welsch: Phys. Rev. A **54**, 1661 (1996)
37. L. Knöll, S. Scheel, E. Schmidt, D.G. Welsch, A.V. Chizhov: Phys. Rev. A **59**, 4716 (1999)
38. G.J. Milburn, T.C. Ralph, A. Gilchrist, A.G. White, W.J. Munro, V. Kendon: in Proceedings of the 6th International Conference on Quantum Communication, Measurement and Computing, Boston, July 2002, eds. J. Shapiro and O. Hirota (Rinton Press, Princeton, 2002)
39. A.M. Steane: Phys. Rev. A **68**, 042322 (2003)

Spin-Based Quantum Dot Quantum Computing

X. Hu

Department of Physics, University at Buffalo, The State University of New York, 239 Fronczak Hall, Buffalo, NY 14260-1500, USA
xhu@buffalo.edu

Abstract. This chapter is an overview of the research on spin-based quantum dot quantum computation, which covers two most prominent architectures, one based on GaAs quantum dots, the other on phosphorus donors in Si. Particular topics include the electron spin coherence in GaAs and Si, the conditions for the Heisenberg Hamiltonian to be a satisfactory description of two-electron interaction in a GaAs double dot, the possibility of using multi-electron quantum dot for quantum computing, the implication of band structure in case of Si, schemes of spin detection, schemes to measure single electron properties in ensemble-averaged experiments, and other related issues. Current status of research in several experimental fronts that are related to quantum dot quantum computing are also discussed.

1 Introduction

During the past decade, the study of quantum computing and quantum information processing has generated wide-spread interest among physicists from areas ranging from atomic physics, optics, to various branches of condensed matter physics [1, 2]. The key thrust behind the rush toward a working quantum computer (QC) was a quantum algorithm designed by a computer scientist, Peter Shor from AT&T, that can factor large numbers exponentially faster than any available classical algorithms [3]. This exponential speedup is due to the intrinsic parallelism in the superposition principle and unitary evolution of quantum mechanics, thus it requires a computer that is made up of quantum mechanical parts (qubits), whose evolution is governed by quantum mechanics. Since the invention of Shor's factoring algorithm, it has also been proved that error correction can be done to a quantum system [4], so that a practical QC does not have to be forever perfect to be useful. After these two important developments, the field of quantum computation has seen an explosive growth.

Many physical systems have been proposed as candidates for qubits in a QC. Prominent examples include trapped ions [5, 6], photons or atoms in cavities [7, 8], nuclear spins in a liquid NMR system [9], electron spin in semiconductor quantum dots [10, 11], donor electron or nuclear spins in semiconductors [12, 13, 14], structures consisting of superconducting Josephson junctions [15], and many more. The experimental demonstration of quantum coherence and maneuverability of these physical systems can be characterized

X. Hu: *Spin-Based Quantum Dot Quantum Computing*, Lect. Notes Phys. **689**, 83–114 (2006)
www.springerlink.com   © Springer-Verlag Berlin Heidelberg 2006

as very fruitful in some cases, such as trapped ions [16] and liquid state NMR [17], to preliminary in many other cases, such as most solid state schemes.

Although experimental progress in many solid state schemes has been slow to come, they are still often considered promising in the long term because of their perceived scalability. After all, the present computer technology is based on semiconductor integrated circuits with ever smaller feature size. However, it is still to be demonstrated whether and how the available semiconductor technology can help scale up an architecture that is quantum mechanically coherent.

Here we first present a brief overview of the current theoretical and experimental progresses in the study of quantum dot-based quantum computing schemes. We then focus on the spin-based varieties, which are generally regarded as the most scalable because of the relatively long coherence times of electron and nuclear spins.

2 General Features
of the Quantum Dot Quantum Computing Schemes

2.1 Classification of the QC Schemes

Many proposals have been made to use quantum dot (QD) and various electron and nuclear degrees of freedom to process quantum information. Crudely these proposals can be classified as charge-based and spin-based (both electron and nuclear spins). We comment on a few of the charged-based proposals below and then discuss in more detail the particular spin-based schemes on which we focus in this lecture.

One of the earliest charge-based proposal is to use the lowest two orbital energy levels of a single electron QD [18]. When an external electric field is applied to the QD, the ground and first excited states would acquire opposite electric dipole moments (the so-called quantum-confined Stark effect [19]). With the interdot tunneling completely suppressed, the inter-qubit coupling is then dominated by dipolar interaction. It creates energy shifts in the energy levels of one QD depending on the state of its neighboring dot, thus provides the physical basis for conditional two-qubit operations [18, 20]. Resonant optical pulses can then be used to implement the conditional excitations.

Several charge-based proposals have since been suggested with more concrete architectures and more detailed physical descriptions. Examples include those using pillars of vertically stacked QDs [21], and chains of horizontal double dot [22, 23]. Similar dipole-interaction-based proposals were also put forward in other physical systems such as trapped neutral atoms [24]. Alternatively, there have also been suggestions using cavity modes (instead of dipole interaction) to couple electronic orbital states [25]. Indeed, the most prominent charge-based qubit is the Cooper pair box proposal in superconductor, where the charged and uncharged state of a small superconducting grain form the basis of a qubit [15, 26, 27, 28, 29]. The semiconductor analogue of the coherent charge oscillation experiment [27] has been done recently [30], generally with shorter coherence time compared to the superconducting counterpart. The great experimental successes in Cooper pair boxes have also prompted searches for other systems that share some common features with Cooper pair boxes. Examples include such exotic systems like quantum Hall bilayers [31, 32].

All the charge-based schemes mentioned so far use singly charged semiconductor QDs. The associated strong Coulomb interaction provides a convenient means for fast qubit manipulation, but can also lead to fast decoherence. One way to alleviate this problem is to use neutral excitations such as excitons as qubits, where there is the added benefit that excitons can be precisely controlled optically. Indeed, uncharged QDs have been proposed as possible candidates for quantum information processing [33, 34, 35, 36, 37], and many experiments have been done to demonstrate exciton coherence and control in a single QD [38, 39, 40, 41, 42, 43]. Here, single excitons are optically excited in individual QDs and can be coherently manipulated optically. The presence and absence of an exciton in a QD provide the two states of a qubit. Entanglement between different qubits is based on Coulomb renormalization of the energy levels. The exciton-based QC proposals clearly illustrate the dichotomy faced by all QC architectures: excitons are neutral, therefore are more insulated from their environment and decohere more slowly than the single charge based schemes. However, the charge neutrality also strongly reduces the interaction between spatially separated excitons, thus rendering it more difficult to perform entangling operations.

In the following we will focus on spin-based QD QC architectures [44]. A fermionic spin, more specifically a spin-1/2, being a quantum two-level system in a finite magnetic field, is a natural qubit with its spin-up and spin-down states. For example, the most successful experimental demonstration of quantum control and entanglement is in a trapped ion system using two hyperfine split nuclear spin levels for the qubit [5]. It was discussed as early as the mid-1990s that Ising interaction between electron spins in solids can produce the desired entanglement for a QC [45]. Here, the specific common thread among the schemes we will concentrate on is that in all of them direct electron exchange coupling plays a crucial role. There are certainly proposals where electron spin interaction takes other forms (such as cavity photon mediated [11], free electron mediated [13, 46], optical RKKY interaction [47], and dipole coupling [48]), and there has been a tremendous amount of research done on the optical characterization of electron and nuclear spins [49, 50, 51, 52]. Nevertheless, we are going to be mostly focused on the electric control of spins and their interaction.

2.2 GaAs Quantum Dot QC Architecture

One of the earliest proposed solid state QC schemes uses the spin of a single electron trapped in a GaAs QD as its qubit (see [10, 53, 54, 55, 56] and references therein, and Fig. 1). Local magnetic fields are used to manipulate single spins, in the sense that it creates a local Zeeman splitting, which can then be accessed by a resonant RF pulse. Inter-dot exchange interaction, which is a spin interaction of electrostatic origin in the form $J\mathbf{S}_1 \cdot \mathbf{S}_2$, is used to couple neighboring spins and introduce two-qubit entanglement. A single trapped electron in a GaAs QD ground orbital state means very low spin-orbit coupling as the electrons occupy states at the bottom of the GaAs conduction band and have essentially S type states [56]. Thus the electron spin coherence time should be even longer than in the bulk, where electron spin decoherence is already very slow.

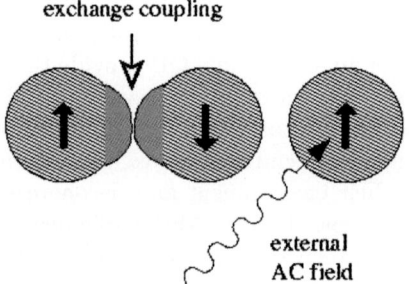

Fig. 1. A schematic of a QD QC

Some of the critical issues regarding GaAs QD QC are trapping a single electron in a gated QD, producing a local electron Zeeman splitting that is different from all its neighbors (so that resonant single spin rotations can be performed), creating and controlling a finite exchange coupling between electrons in neighboring QDs, and last but not least, measuring the single electron spins with high fidelity.

2.3 Si Quantum Dot QC Architecture

The GaAs QD QC architecture we discussed above can be relatively easily extended to a Si/SiGe material system [57]. Here the electrons are confined in the pure (but strained) silicon region (instead of SiGe alloys), while the confinement is produced by the SiGe alloys along the growth direction and surface gates in the in-plane directions (Fig. 1). Compared to GaAs, electron spin decoherence due to hyperfine coupling (as we will discuss in Sect. 3) in such a system can be suppressed by using purified ^{28}Si as the host material. On the other hand, the more complicated band structure of silicon and the less well-controlled interface may pose problems to the coherent manipulation of a quantum device, which need to be further studied.

2.4 Si Donor Nuclear Spin QC Architecture

The QDs underlying the previous two QC schemes are essentially artificial atoms, where electron confinement is provided by the barrier materials and the external electrostatic potential from the metallic gates on the surface of the devices. A naturally occurring alternative is the weakly bound donor states. Take an example of a monovalent donor, where one extra proton is present at the donor nuclear site while an extra donor electron is loosely bound to this proton. Now the donor location becomes the tag for the bound electron spin or the resident nuclear spin, and identical copies of such donors can be easily made in a semiconductor.

One of the most intriguing and influential QC schemes is the donor nuclear spin based Si QC [12], as is shown in Fig. 2a. Here spin-1/2 ^{31}P donor nuclei are qubits, while donor electrons together with external gates provide single-qubit (using external magnetic field) and two-qubit operations (using hyperfine and electron exchange interactions). Specifically, the single donor nuclear spin splitting is given by [12]

$$\hbar\omega_A = 2g_n\mu_n B + 2A + \frac{2A^2}{\mu_B B}, \qquad (1)$$

where g_n is the nuclear spin g-factor (1.13 for ^{31}P [12]), μ_n is the nuclear magneton, A is the strength of the hyperfine coupling between the ^{31}P nucleus and the donor electron spin, and B is the applied magnetic field. It's clear that by changing A one can effectively change the nuclear spin splitting, thus allowing resonant manipulations of individual nuclear spins. If the donor electrons of two nearby donors are allowed to overlap, the interaction part of the spin Hamiltonian for the two electrons and the two nuclei include electron-nuclear hyperfine coupling and electron-electron exchange coupling [12]:

$$\begin{aligned} H &= H_{\text{Zeeman}} + H_{\text{int}} \\ &= H_{\text{Zeeman}} + A_1\mathbf{S}_1 \cdot \mathbf{I}_1 + A_2\mathbf{S}_2 \cdot \mathbf{I}_2 + J\mathbf{S}_1 \cdot \mathbf{S}_2, \end{aligned} \qquad (2)$$

where $\mathbf{S}_1$ and $\mathbf{S}_2$ represent the two electron spins, $\mathbf{I}_1$ and $\mathbf{I}_2$ represent the two nuclear spins, A_1 and A_2 are the hyperfine coupling strength at the two donor sites, and J is the exchange coupling strength between the two donor electrons, which is determined by the overlap of the donor electron wave functions. The lowest order perturbation calculation (assuming $A_1 = A_2 = A$ and J is much smaller than the electron Zeeman splitting) results in an effective exchange coupling between the two nuclei and the coupling strength is

$$J_{nn} = \frac{4A^2 J}{\mu_B B(\mu_B B - 2J)}.\qquad(3)$$

The two donor electrons here are essentially shuttles between different nuclear spin qubits and are controlled by external gate voltages. The final measurement is done by first transferring nuclear spin information into electron spins using hyperfine interaction, then converting electron spin information into charge states such as charge locations [58]. A significant advantage of silicon is that its most abundant isotope ^{28}Si is spinless, thus providing a "quiet" environment for the donor nuclear spin qubits. In addition, Si also has smaller intrinsic spin-orbit coupling than other popular semiconductors such as GaAs. In general, nuclear spins have very long coherence times because they do not strongly couple with their environment, and are thus good candidates for qubits. However, this isolation from the environment also brings with it the baggage that individual nuclear spins are difficult to control and measure. This is why donor electrons play a crucial role in the Si QC scheme. On top of the good material properties Si possesses, there is another potential advantage of a QC based on Si: the prospect of using the vast resources available from the Si-based semiconductor chip industry. In Bruce Kane's own words, it is always advantageous in a fight if one is on the same side as the eight hundred pound gorilla.

2.5 Si Donor Electron Spin QC Architecture

A direct Si donor electron analogue to the GaAs QD QC scheme has also been proposed [14]. In this scheme the ^{31}P donor electron spins are employed as qubits (Fig. 2b). The phosphorus donors are located in a Si layer sandwiched in between SiGe alloy layers. By moving a donor electron into alloy regions of

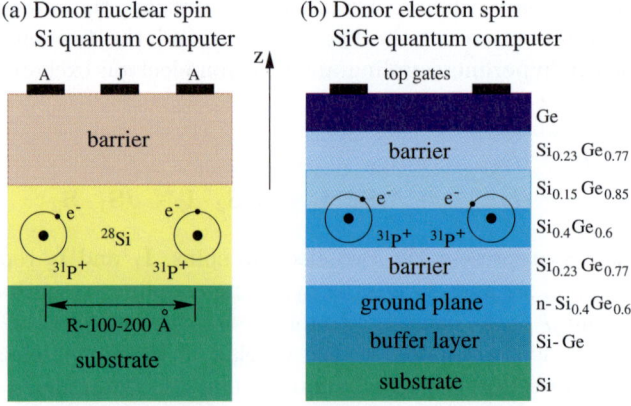

Fig. 2. Schematics of (**a**) a Si donor nuclear spin QC, and (**b**) a Si donor electron spin QC

a different g-factor, its Zeeman splitting can be tuned significantly, which then allows selective single-qubit operations using resonant RF pulses. Similarly, different alloy regions also present different electron effective masses, which affect the size of the donor electron wave function sensitively. Such property can then be used to tune the exchange coupling between two bound donor electrons, and two-qubit operations are again provided by the direct exchange interaction between neighboring donor electrons. This all-electron proposal has much faster gate operations compared to the nuclear-electron hybrid scheme mentioned above. However, the alloy environment has to be more thoroughly studied for such a QC scheme to be considered more practical.

3 Electron Spin Coherence in Semiconductors

To use spins as qubits for a QC, they should possess very long coherence time. Here "long" is in the sense that within a characteristic spin coherence time a large number of operations can be performed. The criteria for the large number are determined by the requirements of successfully performing quantum error correction, which lead to numbers in the range of 10^4 to 10^6 [1]. For example, if a gate operation on an electron spin can be performed in 1 ns, then its coherence time needs to be longer than 10 μs.

Spin coherence (whether it is electron spin or nuclear spin) is regularly described in terms of a longitudinal (or spin-lattice) relaxation time T_1 and a spin-spin relaxation time T_2 [59]. These descriptions originate from the magnetic resonance studies of these spin species going back to the 1940s [60, 61]. T_1 generally describes processes that involve energy transfer between a spin and its environment, while T_2 describes everything that disrupts the quantum coherence of a spin, thus is generally much shorter than T_1.

Notice that spin-flip processes cause both population relaxation and dephasing, contributing to both rates $1/T_1$ and $1/T_2$. However, in a real physical system the longitudinal and transverse directions are often affected differently by the environment. Indeed, there exist pure dephasing processes that affect only T_2 but not T_1. One example is the colliding molecules (which can be described as a pseudospin system, with spin up and down referring to the two relevant internal levels of the molecules) in an optically active gaseous medium, where molecules constantly undergo collisions with each other, some of them inelastic, but most of them elastic. During an inelastic collision, electrons in the molecules undergo transitions that correspond roughly to spin relaxation (or complete loss of the system as an electron gets excited out of the two optically active levels). During an elastic collision, the single molecule energy spectrum changes due to the presence of the other molecule nearby. This shift in energy levels (particularly the two active levels) is dependent on the details of the collision, and is thus a random variable, which we refer to as $\delta\omega(t)$. Including this frequency shift, the differential equation for the off-diagonal density matrix element $\rho_{\uparrow\downarrow}$ (representing the coherence between the up and down levels) becomes [62]

$$\dot{\rho}_{\uparrow\downarrow}(t) = -i[\omega + \delta\omega(t)]\rho_{\uparrow\downarrow}\,. \tag{4}$$

Note that in the first order approximation the population in the up level, $\rho_{\uparrow\uparrow}$, is not affected by this level splitting fluctuation since its equation of motion is independent of the energy difference ω [62] (in other words, the fluctuation in the level splitting does not lead to population relaxation). $\delta\omega(t)$ is a random variable that averages to zero, $\langle\delta\omega(t)\rangle = 0$. Within the Markovian approximation, the random fluctuation in energy level splitting of the two level system causes a pure dephasing effect:

$$\langle\rho_{\uparrow\downarrow}(t)\rangle = \langle\rho_{\uparrow\downarrow}(0)\rangle e^{-i\omega t} e^{-\gamma_{ph}t}\,. \tag{5}$$

This pure dephasing only contributes to T_2 of a spin or pseudospin system, but not to T_1.

Another well-known example of pure dephasing is the dipolar spin-spin interaction between nuclear spins in a solid, which produces effective local magnetic field fluctuations for each nuclear spin and hence contributes essentially only to T_2 (the corresponding effect on T_1 is extremely small). What is important for dephasing is that some change in the state of the environment must occur due to its interaction with the system–dephasing does not require an inelastic scattering process in the system, although all inelastic scatterings necessarily produce dephasing. In fact, as mentioned before, T_2 in the context of electron spin resonance (ESR) and nuclear magnetic resonance (NMR) is often called the spin-spin relaxation time because the most important intrinsic effect contributing to T_2 is the dipolar interaction among various spins in the system, which, while transferring energy among the spins themselves, does not lead to overall energy relaxation from the total spin system (and does not change the total magnetic moment). By contrast, spin-lattice interactions lead to energy relaxation (via spin-flip processes) from the spin system to the lattice, and thus contribute to T_1^{-1}, the spin-lattice relaxation rate. In short, T_2 sets the time scale for the spin system to achieve equilibrium within itself whereas T_1 sets the time scale for the global thermodynamic equilibrium between the spin system and the lattice. For the purpose of quantum computing, it is obvious that T_2 is the directly relevant time scale, because we need to keep all the spins completely quantum coherent.

As we mentioned before, it is imperative that T_2 for an electron spin in a single QD is a factor of 10^4 or so greater than the typical gating time in a QD QC [2] in order for the quantum computing process to be fault tolerant. For $B = 1$ T, the Zeeman splitting in a GaAs QD is about 0.03 meV, which yields 100 ps for the precession time of one spin, and can be used for the one qubit gate (the two qubit gate time is shorter, $\hbar/J \sim 50$ ps for $J \sim 0.1$ meV). Therefore for quantum error correction to be performed reliably, T_2 for the trapped electron spin needs to be on the μs time scale, which may very well be the case at low enough temperatures in a single QD. We note that the existing experimental estimates of free electron spin relaxation time T_1 in GaAs (for $T = 1$–4 K) is around 10–100 ns [63], which is obviously only a lower bound.

So far most studies of qubit decoherence focus on single qubits such as a single electron or nuclear spin. Since universal quantum computation requires two-qubit operations, when two or more qubits are coupled together, it is also of critical importance to study decoherence properties of coupled systems. There have been several theoretical studies along this line. For example, Ref. [64] carefully studied the decoherence during a CNOT gate operation for three different classes of qubits (superconducting flux, charge, and spin) and produced lower bounds of the gate fidelity for these qubits. References [65, 66, 67] explored whether qubit interactions could potentially produce logical qubits with lower decoherence rates. More recently, there has been an experimental study of the relaxation between two-electron singlet and triplet states split by the exchange coupling. It is foreseeable that studies of the decoherence of coupled qubit systems will further expand as they are an integral part of a quantum information processor.

3.1 Spin Decoherence Channels in Semiconductors

In doped semiconductors there are three major spin relaxation mechanisms: Elliot–Yafet (EY), Dyakonov–Perel' (DP), and Bir–Aronov–Pikus (BAP) mechanisms, for conduction electrons [49, 69] at not-too-low temperatures. The origin of the EY mechanism is spin-orbit coupling. Spin-orbit coupling does not lead to spin relaxation by itself, but it mixes the electron orbital and spin degrees of freedom. When combined with another scattering mechanism, such as phonon emission/absorption or impurity scattering, electron spin-flip can occur (spin-flip now means the dominant spin component in the Bloch state changes). In the DP mechanism, the splitting of spin up and down conduction bands due to lack of inversion symmetry (as in III-V semiconductors such as GaAs, which has the zinc-blende lattice structure) acts as an effective momentum dependent magnetic field $\mathbf{B}(\mathbf{k})$. An electron with momentum $\mathbf{k}$ and spin $\mathbf{S}$ precesses in this effective field $\mathbf{B}(\mathbf{k})$ and loses its spin memory. As this electron is scattered into a different $\mathbf{k}$ state, its spin will start to precess around the new effective field. This constant change of effective magnetic field actually reduces the electron spin relaxation, so that the spin relaxation time is inversely proportional to the momentum relaxation time in this mechanism (cf motional narrowing [60, 61]). Lastly, the BAP mechanism is given by the exchange interaction between electrons and holes. Electronic spins move in an effective field produced by the hole spins, and relaxation takes place when hole spins change in a rate much faster than the electron precession frequency.

For the purpose of QD quantum computing, electrons are individually confined in the QDs or around donor nuclei. This quantization of the electron orbital motion should significantly reduce the cross-section of the spin relaxation channels discussed above. However, if the confined electrons are close to an interface, the associated electric field would increase the strength of spin-orbit coupling for such electrons (for example, in GaAs, whose lattice lacks inversion symmetry). Overall, spin-orbit coupling is relatively weak in the regime for quantum computing and is less important than at higher temperatures and electron densities. Furthermore, without scattering off phonons or impurities, spin-orbit coupling can actually be useful rather than harmful to quantum computing [70, 71]. At this limit, other sources of electron spin decoherence, such as electron-nuclear spin coupling, can be more significant, as we will discuss in the next section.

Phonon-assisted spin flip rates due to spin-orbit coupling in a single electron GaAs QD has been calculated [72]. As discussed above, due to wave function localization, the spin orbit relaxation mechanisms for a free electron (EY and DP) are strongly suppressed in a QD, giving a long spin flip time: $T_1 \approx 1$ ms for $B = 1$ T and $T = 0$ K. It was further noticed that spin relaxation is dominated by the EY mechanism, which yields $T_1 \propto B^{-5}$ for transitions between Zeeman sublevels in a one electron QD.

These calculations are consistent with recent transport measurements of spin relaxation in both vertical and horizontal QDs [73, 74, 75]. Pulses of current were injected into a QD coupled to leads in the Coulomb Blockade regime, where the decay rate from excited states can be measured by analyzing the currents generated by the pulses. The results indicate that, for $T = 150$ mK and $B = 0 - 2$ T, spin relaxation times (T_1) are longer than at least a few μs in a many-electron QD (less than 50 electrons), and are longer than 50 μs in single electron dots. This is encouraging from the perspective of the spin-based solid state QC architectures where spin decoherence times of μs or longer are most likely necessary for large scale QC operation.

It is important to keep in mind that T_2 is a more directly relevant decoherence time for quantum computing. Therefore experimental determination of single electron and nuclear spin T_2 in semiconductor nanostructures is crucial, but is still to be performed. ESR combined with transport techniques in principle could be used to probe T_2 in a QD in the Coulomb Blockade regime, just like transport techniques were used to detect ESR in two dimensional electron systems [63]. For example, it was proposed [76] that by applying an AC pump field to a single electron QD subjected to a magnetic field, the stationary current through this QD will exhibit a peak as a function of the pump frequency, whose width will yield a lower bound on T_2.

3.2 Spectral Diffusion for Electron Spins

Spin-orbit coupling does not lead to pure dephasing effect, so that one should expect to have $T_2 = 2T_1$ in materials where spin-orbit coupling in combination with phonon/impurity scattering dominates the spin relaxation, such as high quality GaAs at higher electron density. When the strength of spin-orbit coupling or the cross-section of scattering is reduced, other possible spin decoherence channels have to be considered.

In GaAs, all nuclear isotopes (^{69}Ga, ^{71}Ga, and ^{75}As) have spin 3/2, and there is finite hyperfine interaction between conduction electrons and all these isotopes [77]. The hyperfine interaction is essentially an on-site dipole-dipole interaction between electron and nuclear spins and is $\propto \mathbf{S} \cdot \mathbf{I}$, which includes terms that lead to simultaneous spin flip-flop of both electron and nuclear spins. At a finite magnetic field, the Zeeman splittings of electron and nuclear spins are different by three orders of magnitude, thus energy conservation requires another process to be involved in the transition, therefore reducing the cross-section of such type of processes [78]. However, in the hyperfine interaction there is also a term that is proportional to $S_z I_z$, where the nuclear spins basically produce an effective magnetic field for the electron (assuming one electron trapped in a QD in our situation). If the nuclear spins are all frozen in their respective states, they would simply produce a random but fixed field, which would result in the so-called inhomogeneous broadening for the electron and the effect can be accounted for by calibration and spin echo techniques [60, 61]. If the nuclear spins are dynamically coupled (through dipolar coupling, for example), though, the trapped electron would be in a magnetic field that fluctuates both spatially and temporally. This fluctuating field makes the electron Zeeman splitting a random variable that undergoes the so-called spectral diffusion, which results in pure dephasing for the electron spin [79]. The calculated electron spin dephasing time for GaAs QDs is in the order of tens of μs.

Theoretically, it can be envisioned that if all the nuclear spins are polarized, the corresponding electron spin spectral diffusion and dephasing can be suppressed [55]. However, 100% polarized nuclear spins would create a significant effective magnetic field, which can have its own negative side effect. Furthermore, the creation of this high degree of polarization is also nontrivial. In other words, ingenious approaches need to be devised to deal with the nuclear spins in a GaAs QD, and we may not have seen the optimal approach yet.

Spectral diffusion can occur to donor electrons in Si as well, when the host Si material is not purified: Naturally occurring Si contains more than 95% of ^{28}Si and ^{30}Si, which have no nuclear spin, and nearly 5% ^{29}Si, which has nuclear spin 1/2. For a confined donor electron (at a phosphorus site, for example), which has a Bohr radius of about 3 nm, there can be many nuclear spin-1/2s (more than 10^2 of them within a sphere of radius of 30 nm) with finite hyperfine coupling to the electron, and thus can produce spectral diffusion in this electron as discussed above [79]. Fortunately, in Si there is a way to reduce the effect from the nuclear spins of ^{29}Si: isotopic purification. Again, if a complete purification can be achieved, the nuclear spin induced electron spin spectral diffusion can be completely suppressed [79].

4 Spin Manipulations and Exchange

4.1 Spin Hamiltonian in a GaAs Double Quantum Dot: Coulomb Interaction and Pauli Principle

One of the key issues in the spin-based QD QC involving exchange coupling in a coupled QD is to accurately describe the electron interaction in terms of a spin Hamiltonian such as a Heisenberg exchange Hamiltonian with appropriate correction terms. Ideally, for small QDs at low temperatures, the electron orbital degrees of freedom are frozen, so that the only states two electrons can possibly occupy in a double dot are the ground spin singlet and triplet states, whose splitting is the exchange splitting J. However, it is important to clarify how well the ground state manifold is separated from the excited states, and whether the exchange splitting J in the ground state manifold is sufficiently large to support a practical QC.

The Hamiltonian for two electrons in an electrostatic confinement produced by surface gates and growth direction barriers can be written as

$$H = \sum_{i=1,2} \left[\frac{1}{2m^*} \left(\mathbf{p} + \frac{e}{c} \mathbf{A}(\mathbf{r}_i) \right)^2 + V(\mathbf{r}_i) \right] + \frac{e^2}{\epsilon r_{12}} + \sum_{i=1,2} g^* \mu_B \mathbf{B}(\mathbf{r}_i) \cdot \mathbf{S}_i . \quad (6)$$

This is an effective mass Hamiltonian where the underlying Bloch function at Γ point (the bottom of the GaAs conduction band) is already factored out. Notice that none of the spin-dependent terms except Zeeman coupling are included in this Hamiltonian. In essence, electrostatic interaction is much stronger than direct magnetic interactions such as magnetic dipole interaction. Neglected interactions include electron spin-orbit interaction (which will be discussed later on in this subsection), electron-electron magnetic dipole interaction, electron-nuclear spin contact hyperfine interaction (discussed in the previous section on electron spin spectral diffusion), electron-nuclear spin dipole interaction, and other higher order interactions. In fact, in arriving at the effective mass Hamiltonian (6), small spin-mixing terms inversely proportional to the conduction band gap are also neglected [80].

Here we have separated the two-electron-interaction related terms of the Hamiltonian from those that involve interactions between either or both of the electrons and the surrounding environments that include the crystal lattice (in terms of phonons), the nuclear spins of the ions, and other electrons present in the system. The effects of these interactions are categorized as decoherence, in the sense that as soon as electron spin coherence passes into these channels, the chance of a revival of the coherence is vanishingly small.

The theoretical calculation of electron spin exchange in a double dot was first done using the Heitler-London approach, in which the electron orbital states are limited to the ground states in the two single QDs that form the double dot [55], and double occupied single dot states are excluded. The effects of the lowest two double-occupied states (where both electrons are in the same dot) are included in the so-called Hund-Mulliken calculation [55]. Both calculations indicate that the exchange constant can be sizable in a GaAs QD system.

More accurate calculations of the exchange coupling and the overall spectrum of the two QD-confined electrons can be performed with larger basis of single dot states. For example, we performed a molecular orbital calculation to further clarify the properties of the exchange splitting in a GaAs horizontal double QD by including the excited P orbital states of the two QDs [56]. The inclusion of the anisotropic P orbitals provides more flexibility to electron wave function deformation and bonding, and thus leads to more faithful description of the exchange splitting of the two-electron ground states. After all, molecular bonding is strongly affected by electron distribution in space, while the system we considered is essentially an effective two-dimensional hydrogen molecule. The double dot we stuidied is formed from two-dimensional electron gas by surface gate depletion (the so-called horizontal or lateral QDs). The growth direction confinement (due to AlGaAs alloys) is so strong that the excitation energy scale along that direction is much higher than the horizontal direction and excitation along that direction is neglected (thus the name horizontal or lateral QD). One important advantage of the horizontally coupled QDs is that the inter-dot coupling can be easily tuned with surface gate potential adjustments. Our numerical results showed that the inclusion of the P orbitals indeed affect the exchange coupling quite significantly, generally causing an increase of about 20% compared to the Hund-Mulliken calculation [56] (see Figs. 3 and 4). The size of the QDs we considered is quite small (in the order of 40 to 50 nm in diameter), somewhat smaller than the state of the art experimental value of about 100 nm. The increase in size invariably leads to smaller on-site Coulomb repulsion, smaller single particle excitation energy, and generally smaller exchange coupling.

The spectroscopic results for double QD [55, 56] show that exchange coupling should be sufficiently strong to implement quantum computing operations, where a basic controlled-NOT operation can be built from single qubit spin rotations and two so-called square-root-of-swap gates [10]:

$$U_{\mathrm{CNOT}} = e^{i\frac{\pi}{4}\sigma_{2y}} \, e^{i\frac{\pi}{4}\sigma_{1z}} \, e^{-i\frac{\pi}{4}\sigma_{2z}} \, U_{\mathrm{sw}}^{\frac{1}{2}} \, e^{i\frac{\pi}{2}\sigma_{1z}} \, U_{\mathrm{sw}}^{\frac{1}{2}} \, e^{-i\frac{\pi}{4}\sigma_{2y}} \,, \qquad (7)$$

where σ are Pauli matrices, and $U_{\mathrm{sw}} = \exp(i\pi\sigma_1 \cdot \sigma_2)$. For example, an exchange splitting of 0.1 meV corresponds to a swap gate with duration as short as 100 ps. Compared to the electron spin decoherence time that might be as long as microseconds, such gating time is sufficiently short for quantum error correction codes to work properly.

By including only the lowest spin singlet and triplet states, we have implicitly assumed that the higher energy two-electron states can be neglected. Such an assumption is based on the electron states being manipulated adiabatically. On the other hand, the gate operations are limited in duration by the electron spin decoherence time. It is thus imperative to determine what

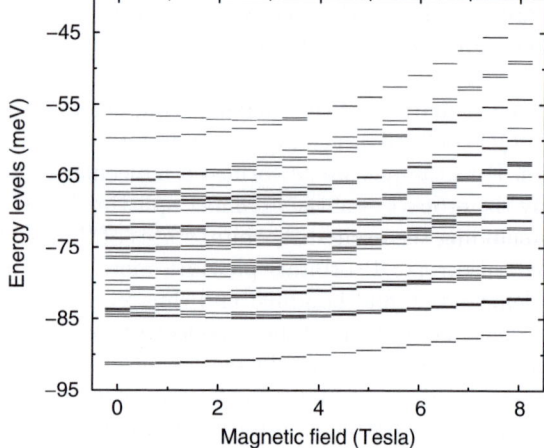

Fig. 3. Two-electron energy spectrum of a GaAs horizontal double QD [56] as a function of applied perpendicular magnetic field. The ground singlet-triplet manifold is well separated from the excited states. Here the inter-dot distance is 30 nm, and the ground electron wave function radius is about 10 nm

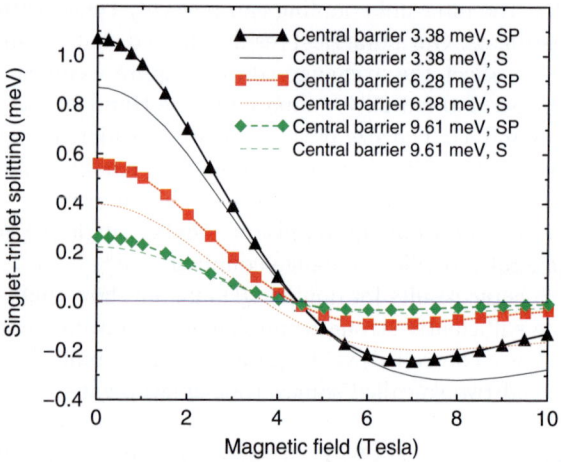

Fig. 4. Exchange splitting between the ground singlet and unpolarized triplet states of a GaAs horizontal double QD [56]

the adiabatic requirement is in the present architecture. Such a calculation was carried out within the Hund-Mulliken model by assuming a particular temporal shape of the exchange splitting J in a double QD [81]. We performed such a calculation with calculated two-electron energy spectra and wave functions [82]. Instead of assuming a particular temporal profile for the exchange coupling J, we assumed a temporal variation in the inter-dot barrier height, which is a quantity that can be directly tuned by external voltages applied to the surface gates. Our results show that errors due to non-adiabaticity decreases rapidly as the gate operations become longer, and the requirement for adiabatic operation is not overbearing on the spin-based GaAs QD QC [82].

The calculations on the energy spectra and adiabatic manipulation thus justify the use of Heisenberg spin exchange Hamiltonian to describe the low energy dynamics in a two-electron double QD and the related two-qubit operations in a spin-based QD QC. In these calculations spin-orbit coupling has been neglected because of their small magnitude for the conduction electrons in bulk GaAs (near the bottom of the conduction band the electron wave function is mostly formed from the atomic S orbitals). However, the horizontal QDs are made from two-dimensional electron gas confined in heterostructures or quantum wells, where the sharp surfaces and asymmetry lead to strong Rashba type spin-orbit coupling (even for symmetric quantum wells, this spin-orbit coupling does not vanish because of the lack of inversion symmetry in GaAs), which in turn leads to finite anisotropic exchange in the spin interaction [83] in the form of $\mathbf{h} \cdot (\mathbf{S}_1 \times \mathbf{S}_2)$ plus higher order corrections. Here $\mathbf{h}$ is a vector determined by spin-orbit interaction. The inclusion of these terms does not take away the capability of QDs to perform quantum logic operations (indeed, there has been a proposal to use the anisotropic exchange coupling to perform quantum logic operations [70, 71]). Nevertheless, they do add more complexity to the simple isotropic Heisenberg exchange coupling. Fortunately, it has been proved that by carefully choosing the temporal profile of the inter-dot coupling (basically maintaining time reversal symmetry), it is possible to largely eliminate the effects of the anisotropic exchange due to the spin-orbit interaction [84].

Even when the two-electron interaction in a double dot can be characterized by the Heisenberg spin exchange Hamiltonian, inhomogeneity in the single spin environment can still cause problems in the two-qubit quantum logic operations. For example, we showed that inhomogeneous Zeeman coupling leads to incomplete swap operations [85]. This means that swap cannot be accomplished by a single pulse of exchange gate anymore. Instead, several pulses (at least 3) have to be used for large inhomogeneity [86], while smaller inhomogeneities such as those due to trapped charges nearby can be and have to be corrected [85]. Interestingly, inhomogeneous Zeeman coupling can also be utilized for the purpose of qubit encoding [87], which leads to all-exchange logical operations (eliminating the need for local magnetic field and/or g-factor engineering) [88] that originate from the concept of decoherence free subspace [89].

To relax the requirement of spin-based QD quantum computation, multi-electron QDs have also been studied as candidates for qubits [90]. For instance, we performed a configuration interaction calculation for a double dot with six electrons (three per dot) to explore whether the low-energy dynamics is still entirely dominated by spin dynamics. Our results showed that the ground state complex is well separated from the higher energy excited states (Fig. 5) at relatively high magnetic fields, and the splitting between the lowest-energy singlet and triplet states can be sizable as well (Fig. 6). However, our results also showed that orbital level degeneracy can lead to the participation of multiple states in the low energy dynamics at zero or low magnetic fields, therefore causing serious complexity and difficulty in spin exchange. To solve this problem, external means such magnetic field or quantum dot deformation has to be applied to lift the orbital degeneracy so that the electron cloud in each QD can again be described by an effective spin-1/2 entity. Another theoretical study has also clearly demonstrated a variety of difficulties in the control of gate operations when one attempts to use multi-electron QDs as qubits [91].

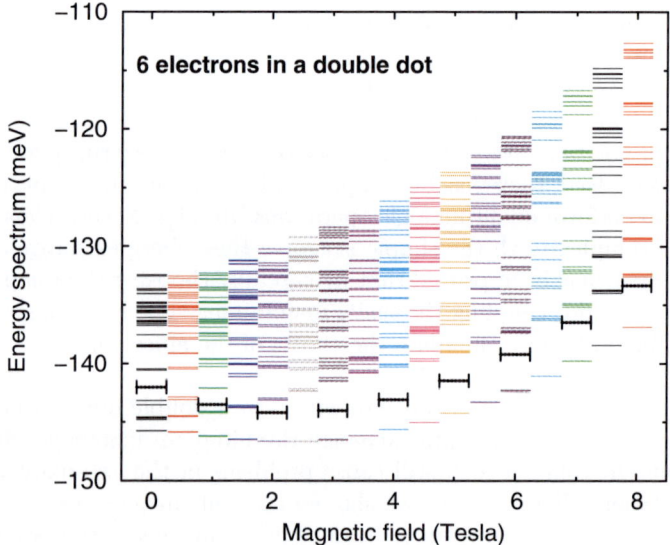

Fig. 5. Six-electron energy spectrum of a GaAs horizontal double QD [90]

In using a tunable exchange interaction for quantum gates, one needs to control the magnitude and duration of the interaction precisely. Since exchange interaction depends sensitively on the wave function overlap, its precise control is of critical importance. One suggestion on how a good control can be achieved is to control the way the wave function overlap is tuned by altering the geometry of the QD designs and is called a pseudo-digital approach [92]. As the experimental study of QD QC pushes toward two and more qubits and qubit manipulations, such considerations are directly relevant to the designs of next generation architectures and ultimately the scale-up of any of the QD-based QC schemes.

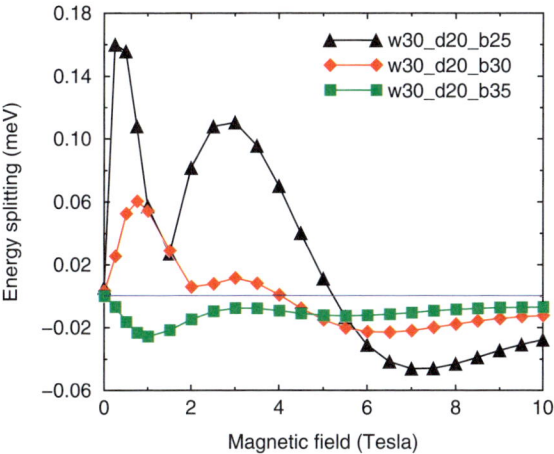

Fig. 6. Energy splitting of the two lowest energy states of a six-electron GaAs horizontal double dot [90]

4.2 Implications of Si Conduction Band Structure to Electron Exchange

As we mentioned before, Si possesses a variety of nice material properties (small spin-orbit coupling, spinless isotopes, etc.) for the purpose of quantum computing, so that it is clearly one of the favored host materials for a solid-state QC. However, Si does have one complexity that GaAs, the other popular semiconductor material, does not have: The Silicon conduction band has six minima close to the X points of the silicon First Brillouin Zone [93], so that donor electron wave functions have to be expanded on the basis of the six Bloch functions at these points. It was pointed out in the context of donor magnetic phase transition that the presence of degenerate conduction valleys leads to valley interferences and a shift to smaller values for average electron exchange coupling [94]. The potential problem such interference effects may cause to a donor based Si QC was also mentioned in the original proposal [12].

To quantitatively address this concern over donor exchange coupling in Si, which is a crucial link for two-qubit operations in the Si QC architecture, we have performed a series of Heitler-London type calculations of the donor electron exchange coupling. Such a calculation is based upon the single donor wave functions, which can be expressed as [93, 94, 95, 96]

$$\psi(\mathbf{r}) = \sum_{\mu=1}^{6} \alpha_\mu F_\mu(\mathbf{r})\phi_\mu(\mathbf{r}) = \sum_{\mu=1}^{6} \alpha_\mu \mathbf{F}_\mu(\mathbf{r})\mathbf{u}_\mu(\mathbf{r})e^{i\mathbf{k}_\mu \cdot \mathbf{r}}, \qquad (8)$$

where F_μ are the so-called envelope functions while $\phi_\mu(\mathbf{r})$ are the Bloch functions at the bottoms of the Si conduction band [95]. $|\mathbf{k}_\mu| \approx 0.85 \cdot 2\pi/a$ is the location of the conduction band minima and is very close to the X points. The presence of these plane wave phase factors leads to a significantly more complicated expression (compared to, for example, GaAs) for the electron exchange splitting [96]:

$$J(\mathbf{R}) = \sum_{\mu,\nu} |\alpha_\mu|^2 \, |\alpha_\nu|^2 \mathcal{J}_{\mu\nu}(\mathbf{R}) \cos(\mathbf{k}_\mu - \mathbf{k}_\nu) \cdot \mathbf{R} \, . \qquad (9)$$

Here $\mathcal{J}$ represents integrals over the envelope functions and is thus a slowly varying function of donor positions. The key fact here is that the $\mathbf{R}$-dependence of $J(\mathbf{R})$ is strongly oscillatory because of the sinusoidal factors $\cos(\mathbf{k}_\mu - \mathbf{k}_\nu) \cdot \mathbf{R}$.

Our numerical results, as shown in Figs. 7 and 8, demonstrated that the inter-valley interference indeed leads to strong atomic scale oscillations in the inter-donor electron exchange, which potentially presents a significant difficulty in the control of two-qubit operations [95]. Uniaxial strain can be used to break the Si lattice symmetry and partially lift the degeneracy between the valleys, so that as few as two valleys make up the bottom of the conduction band. Then the sum over μ in (9) is much simplified, but the sinusoidal factor will still remain, so that care still has to be taken in controlling the donor exchange (Figs. 9 and 10) [96]. These results have been corroborated by another calculation that also considered higher order corrections in Coulomb interaction energy [97]. More recently we have attempted to relax the Heitler–London approximation to minimize the two-electron energy [98]. However, the results showed that in the cases of donors, which have very low single particle potential energy near the donor nuclei, the two-electron contribution to total energy is completely dominated by the single particle contributions, thus the Heitler–London results based on single donor states are quite robust [98].

As we mentioned before, there have also been proposals using electron spins in Si or SiGe host materials for quantum computing. The problem with oscillatory exchange also plagues the donor-electron-based scheme. In the case of QDs in Si, the problem becomes more subtle. In such a system there does not exist a strong scattering center such as a donor, so that the different valleys in Si do not couple significantly. A weak inter-valley coupling would bring into question the validity of (8), with its strongly pinned phase for each valley. The valley degeneracy is lifted by the quantum well for the QDs, but the energy splitting is quite small [99]. More theoretical and experimental studies are needed to further clarify this situation.

One way to avoid the potential problems associated with the control of inter-donor exchange coupling is to move a donor electron around between ionized nuclei [100, 101]. Here nuclear spins are still the qubits, while a *single electron* is used as a shuttle to physically move among nuclei to enable effective two-qubit coupling and operations. Since on-site hyperfine coupling has been shown to be quite robust against Si band structure complexities [102, 103], this approach thus maintains the long coherence time advantage

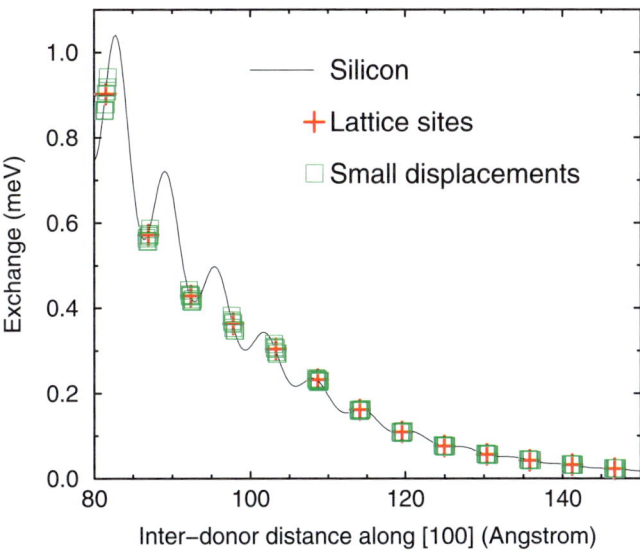

Fig. 7. Donor electron exchange splitting in relaxed bulk Si. The two ^{31}P donors are aligned along the [100] direction [95]

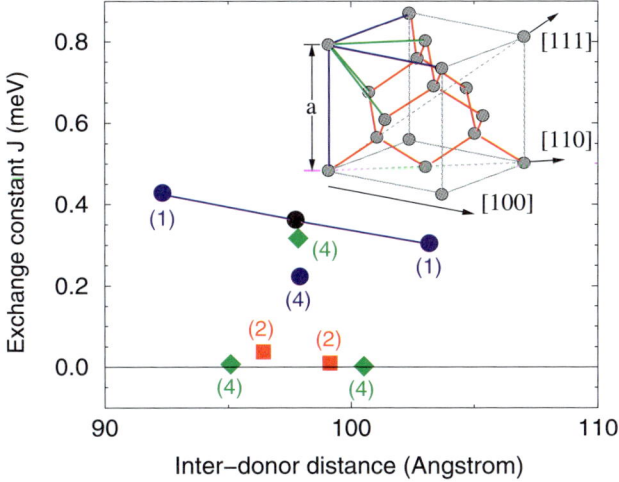

Fig. 8. Donor electron exchange splitting in bulk Si. The two ^{31}P donors are almost aligned along the [100] direction, separated by about 18 lattice constants, and with one of the donors displaced into its nearest neighbor lattice sites [95]

of a donor nuclear spin in Si while completely removes the requirement of inter-donor exchange coupling. However, a potential disadvantage of such a charge-coupled device may be associated with the charge movement and the corresponding spin decoherence. Further analysis is still needed to clarify the physical picture here.

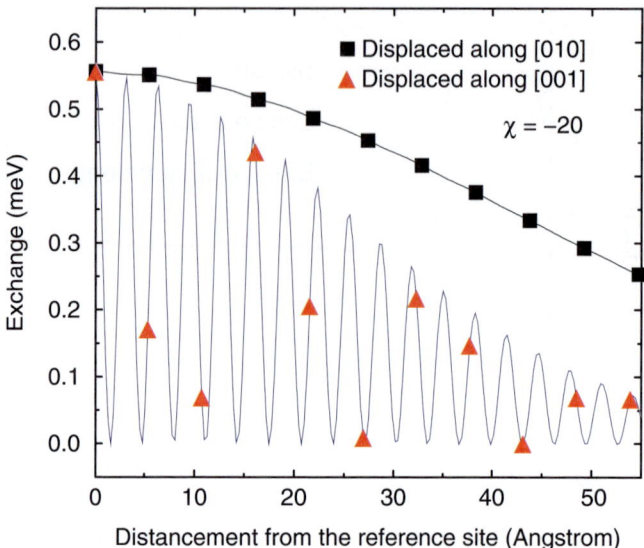

Fig. 9. Donor electron exchange splitting in Si uniaxially strained along [001] direction ($\chi = -20$). The two donors are approximately aligned along the [100] direction, with one of them displaced along the [010] direction [96]

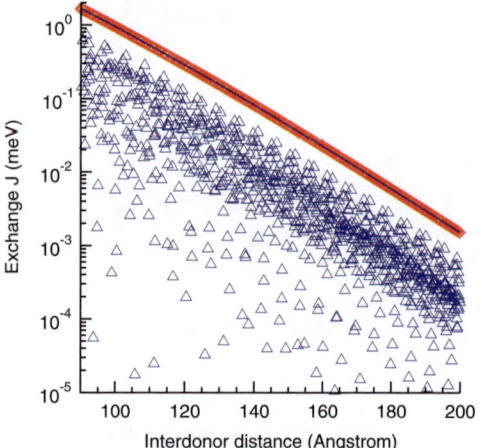

Fig. 10. Donor electron exchange splitting in both relaxed bulk Si and Si strained along the [001] direction. The two donors are both in the (001), or xy, plane. We consider a situation where one of the donors is located in any of the possible lattice positions between two concentric circles of radii 90 Å and 180 Å with the other donor positioned at the center of the circles. The data points correspond to the exchange calculated at all relative positions considered. The solid line is $J(\mathbf{R})$ for $\mathbf{R}$ along the [100] direction for $\chi = -20$ [96]

4.3 Single Spin Detection Schemes

Single spin detection is crucial for any spin-based QC architecture to work properly, both in terms of quantum error correction, and in terms of reading out the final results of a calculation. However, since magnetic force is weak (compared to electrical force), and a single Bohr magneton is a very small magnetic moment, direct measurement within a short duration (such as using the most sensitive magnetometer available at present, a SQUID magnetometer) is almost impossible. However, several techniques are being actively studied and have produced some promising progresses.

One approach to single spin detection is to first convert the electron spin states into electron charge states, for example through spin blockade effects [104, 105, 106, 107]. It is well established now that single electron transistors (SET) and quantum point contacts (QPC) are extremely sensitive charge detectors [108]. Thus, if one can establish a correlation between electron charge location and spin states, a means to determine spin states can be established by observing the current or conductance of the SET/QPC charge detector. One shortcoming of the conventional DC-biased SETs and QPCs is that they are relatively slow detectors, with bandwidth in the order of 10 kHz. However, an alternative approach to DC-SET has been proposed [109]. Instead of measuring current directly, here a pulsed radio-frequency field is sent into a circuit containing an SET and the circuit response is measured. The relevant physical quantity in such a measurement is conductance, which can be determined without collecting a large number of electrons. The bandwidth of these so-called Radio Frequency SETs (RF-SET) has been shown to reach above 10 MHz, so that they are very promising candidates for quantum detectors. Now, if spin states can be efficiently converted to charge locations, single spin detection also becomes possible. One of the early examples of such conversion was suggested in connection with the donor in Si architecture, where the on-site exchange splitting between a singlet and a triplet state on a double valent donor, together with an SET charge detector, is used for spin detection [58]. A similar scheme can also be constructed for artificial QD-based structures.

Quantum measurement of hyperfine split nuclear spin levels in trapped ion systems has been achieved using the so-called quantum jump technique, in which an ancilla electron orbital level and light absorption/emission between this level and the qubit levels are used to read out the quantum state of the qubit with basically 100% efficiency. Such ancilla levels certainly also exist for semiconductor QDs, so that similar quantum jump processes have been suggested for single electron spin detection [48]. Concrete examples of such a scheme have been described and simulated for an electron spin in a QD [110, 111].

In the original proposal for QD QC in GaAs, the so-called spin valve effect is suggested for spin measurement [10], where electron tunneling into another QD depends on its spin orientation, in analogy to giant magnetic resistance [112]. An alternative is to prepare a supercooled paramagnetic dot, so that when an electron tunnels into this QD, a large ferromagnetic domain could quickly nucleate along the incoming spin direction. This more macroscopic magnetization can then be detected in a more traditional way [10]. Furthermore, spin filtering is not limited to using magnetic quantum dots. Regular quantum dot, assisted by Coulomb blockade and Pauli principle, can also work as a spin filter [113, 114].

Another approach toward spin measurement is a direct magnetic force detection using the so-called magnetic resonance force microscopy (MRFM) [115]. In MRFM a small magnet at the tip of a cantilever creates a strongly inhomogeneous magnetic field near the surface of a sample containing paramagnetic electron spins. When a RF field with a certain frequency is applied, only those spins that are Zeeman split by the right amount can get into resonance with the external RF field, so that these spins can apply controlled forces on the cantilever. By measuring the motion of the cantilever, one can infer information of those electron spins that are on resonance (thus the name MRFM).

Single spin detection is not needed for the purpose of characterizing a single or double QD/donor system. For example, we proposed a scheme to use resonant micro-Raman scattering to measure inter-donor exchange coupling [116]. The polarization- and temperature-dependence of Raman scattering provides abundant information of a two-donor system, while resonant photons enhances the cross-section of such scattering process so that single pairs of donors can be observable. Similarly, transport and/or optical techniques can be used to measure the ground exchange splitting in a double QD as well [117].

4.4 Approaches to Generate and Detect Electron Spin Entanglement in Quantum Dots

Two basic ingredients of spin-based QD QC are single spin manipulation and particularly single spin detection. These are very hard tasks and are attracting plenty of attention from experimentalists and theorists alike. In the meantime, traditional ensemble-averaged experiments can also be used to demonstrate quantum coherent properties of electron spins. For example, electron spin entanglement has never been experimentally demonstrated in condensed matter systems in terms of measured correlations, because of the usual presence of strong interaction and the associated difficulty in isolating the target electrons. However, with the help of QDs and transport through them, it is possible to generate and detect electron spin entanglement. From the perspective of QD QC, a reliable source of spin entangled electrons is very important to tasks such as error correction.

Many approaches for creating/detecting entanglement in solid state systems have been theoretically proposed in the literature. For example, Cooper pairs in a superconductor are (usually) in a spin singlet (and entangled) state, thus it is quite natural to consider extracting them with control to make up a source of pairs of spin-entangled electrons [118, 119, 120, 121, 122, 123]. The key here is to separate the two electrons into different drain electrodes using, for example, Coulomb blockade in quantum dots or wires. There are many other physical systems within solid state that have electron spin entangled states as part of their eigenstate spectrum and have been suggested as sources of entangled electrons. Examples include quantum dot and quantum wire electron singlet states in semiconductors [124, 125, 126, 127, 128]. Here we briefly discuss our idea of using a double dot to create spin entangled electron pairs [127].

As discussed in the previous sections, at zero or low magnetic field, the ground state of a two-electron double dot is a spin singlet state, where the spins of the two electrons are entangled $((|\uparrow\downarrow\rangle - |\downarrow\uparrow\rangle)/\sqrt{2})$. However, sequential tunneling through a double dot is generally dominated by the lowest order processes, so that the current through a double dot is generally made up of mostly single electrons tunneling out of the entangled state. To extract pairs of spin-entangled electrons, a relatively straightforward approach is to introduce time-dependent (specifically, periodic) tunnel barriers, so that during part of one cycle the electrons form molecular states (such as the spin singlet state) in the double dot, while during the rest of the cycle the electrons are emptied out into the drain electrodes. Such time-dependent manipulation of tunnel barriers is quite well-understood for single quantum dots, and in principle feasible for a double dot. We estimated various parameters for such a double dot turnstile and found that they are quite reasonable and fall within the capability range of currently available technology.

A possible method to observe solid state electronic entanglement within ensemble-averaged approach is to use two-electron interference in transport. One example is to use a beam splitter [127, 129, 130], into which pairs of entangled electrons are injected. Although this detection scheme is not a true Bell-type measurement, it is nonetheless an important first step as it deals with correlations between electrons that have been extracted from their entangler and separated.

5 Current Experimental Status

5.1 Single Electron Trapping in Horizontal QDs

Since the invention of gated semiconductor QDs, there has been a continuous trend to fabricate structures that can hold smaller and smaller number of electrons. Single electron QD through electrical transport was achieved in vertically contacted InGaAs QDs several years ago [131]. During the past year it has been shown that these vertical dots can be coupled horizontally by growing them next to each other and fabricating additional side gates [132]. For gated horizontal dots, however, the confinement is given by an electrostatic field produced by metallic gates that are 50 to 100 nm away, and the feature width of a metallic gate is generally above 20 nm, so that horizontal QDs made from depleted two-dimensional electron gas are generally large in size but energetically shallow. In these QDs it is difficult to simultaneously increase depletion (to reduce number of trapped electrons) and maintain a finite coupling to the source and drain leads (so that transport characterization can be done). A solution to this problem was found several years ago, when a new design with source and drain coupling controlled by a single gate was studied [133]. The design has since been used to create single electron QDs and one- or two-electron double QDs successfully [68, 134, 135, 136]. Furthermore, recent experiments have shown that electrons trapped in such a QD have significant (and consistent with expectation) Zeeman splitting, and that spin-flip time is extremely long in these confined states [75].

5.2 Single Spin Detection

Although many theoretical proposals have been published on how to achieve single spin detection, experimental demonstration remains a challenging and hotly pursued goal, particularly for solid state spin-based QC architectures, though impressive progress has been made along a variety of lines of research.

In the past two years, QPCs have been used to successfully measure single QD properties such as QD charging and excited state spectroscopy [135, 137]. The simpler structure of a QPC makes it an enticing alternative to an SET detector, and may be easier to integrate into larger structures. Recently, a QPC detector combined with a pulsed-gate technique has produced the first single spin detection in a single quantum dot [138], while a follow-up experiment showed that differentiating tunneling rates can also be used to detect single spins in single-shot measurements [139]. Here a single quantum dot is coupled to a single lead and is being constantly detected by a QPC. By pulsing the gate voltage applied to the quantum dot, it can be populated by a single electron. If an external magnetic field is applied, the spin up and down states will have different energies, which then allows a state-selective emptying of the dot when the gate voltage is adjusted in between the two Zeeman split levels. Since the QPC detector is very sensitive to the presence of an electron in the quantum dot, it can tell whether an electron has emptied out of the dot or not, with a detection bandwidth in the order of 10 kHz (in other words, faster processes cannot be detected). More recently, this combination of pulsed gate and QPC detection has also produced the first measurement of two-spin decoherence in a horizontal double dot at low magnetic fields [68], where electron-nuclear spin hyperfine interaction dominates spin dephasing [140].

In analogy to the quantum jump technique in trapped ion QC schemes [6], optical transitions have been used to observe single spin states of nitrogen-vacancy defect centers in diamond [141]. These observations clearly demonstrated that although solid state systems generally have much more complexity compared to atomic systems, many of the techniques and concepts can be transferred with truly fruitful consequences.

It has recently been reported that single electron spins can be reliably detected with a particular realization of the MRFM [142]. Spin diffusion suppression, which is intimately related to the interaction between the cantilever and the spin being observed, has also been characterized in the inhomogeneous magnetic field produced by the small magnet at the end of the cantilever of an MRFM [143].

5.3 Electron-Nuclear Spin Interaction in QDs

As we mentioned previously, electron-nuclear spin hyperfine coupling can lead to electron spin spectral diffusion and pure dephasing. Thus it is an important environmental issue for electron spins, especially for QD host materials that have a lot of nuclear spins, such as GaAs and InGaAs (the preferred materials for gated horizontal and vertical QDs, respectively).

There have been quite extensive experimental studies of electron-nuclear spin coupling in bulk semiconductors and more recently semiconductor heterostructures [49, 144, 145, 146]. Dynamical nuclear spin polarization has recently been demonstrated in a spin blocked semiconductor double QD made of $Ga_{0.95}In_{0.05}As$ [106, 117, 146]. Coulomb interaction and Pauli principle means that in a double QD two-electron singlet and triplet states are split by the exchange interaction, so that proper voltage offset and bias between the double dot leads to significant suppression in the tunnel current due to occupied triplet state (thus the so-called spin blockade regime, which has been suggested for single spin detection, a crucial component of quantum information processing) [106]. One way to lift this blockade is to apply an appropriate magnetic field, so that singlet-triplet spin-flip transition can be facilitated by the electron-nuclear spin hyperfine coupling as one of the polarized triplet state is energetically degenerate with the singlet state [146]. This selective transition can then dynamically polarize the nuclear spins in the double QD. Indeed, strong experimental evidence has been observed that nuclear spins are indeed being polarized [146], though many physical issues still have to be sorted out before a more thorough understanding of the coupled electron-nuclear spin system can be achieved.

Electron-nuclear spin interaction has also been extensively investigated recently in a series of experiments for horizontally coupled gated double dots [68, 147]. In these experiments it was shown that nuclear spins cause an inhomogeneous broadening of the electron spins and lead to a very short 10 ns electron spin dephasing time, although spin relaxation time at a small field (100 mT) is up to 70 μs. Observations of spin echo and Rabi oscillation shows that the electron spin T_2 time (before inhomogeneous broadening) is above 1 μs, and further experimental tests are still ongoing [147].

5.4 Fabrication of Donor Arrays in Si

Currently there are active experimental research efforts in attempting to fabricate well-controlled ^{31}P donor arrays in a Si crystal, and plenty of experimental progress has been made [148]. Two experimental approaches are adopted, attacking the problem from opposite directions [148, 149]. One uses ion implantation by bombarding ^{31}P ions into crystalline silicon, thus it is also called a top-down approach. The other uses MBE to grow the system layer by layer and STM to identify donor locations, and is called the bottom-up approach.

In the top-down approach, annealing is needed after the bombardment to make the ^{31}P donors substitutional so that they become shallow donors. If they remain interstitial, they would behave as deep centers [150], which have different electronic structures and thus are not useful for the purpose of quantum computing within the Kane proposal. Single electron transistors are used to monitor the presence of donors since they are very sensitive to net charges [148]. At present the precise positioning of the donors and annealing of the Si host lattice are being actively studied [151, 152].

In the bottom-up approach, a clean Si surface is first hydrogenated. An STM is then used to pick off hydrogen atoms at desired locations, after which the surface is put in contact with PH_3 gas so that phosphorus atoms would tend to attach to the surface at the vacancies. This way an ordered array of donors can be fabricated with a high degree of regularity. Using this approach, it has been shown that precise positioning of phosphorus donors into a linear array can be achieved on a Si [001] surface [153]. More recently, the incorporation of the donors into the bulk of Si has also been demonstrated [154]. Although much more experimental efforts are required, and most probably more sophisticated technologies in donor positioning and manipulation have to be invented for this QC architecture to be fabricated with precision, recent experimental progresses are nonetheless quite impressive and promising.

6 Summary

We have presented a brief review of the theoretical and experimental progresses related to spin-based QD QC architectures. We introduced the most prominent proposals of QD QC, outlined the basics on how these proposals might work, explored potential problems with the different material systems, and finally discussed where the present experimental studies stand. We thank US ARO, ARDA, and LPS for continued financial support to our research. The results presented here are the products of collaborations with S. Das Sarma, B. Koiller, D. Drew, R. de Sousa, R.B. Capaz, and fruitful discussions with J. Fabian, I. Žutić, A. Kaminski, R.A. Webb, Y.Z. Chen, B.E. Kane, D. Romero, M. Friesen, E. Mucciolo, S. Tarucha, K. Ono, R. Hanson, and R. Budakian.

References

1. D. DiVincenzo: Science **270**, 255 (1995); A. Ekert, R Jozsa: Rev. Mod. Phys. **68**, 733 (1996); A. Steane: Rep. Prog. Phys. **61**, 117 (1998); A. Galindo, M.A. Martin-Delgado: Rev. Mod. Phys. **74**, 347 (2002); I. Žutić, J. Fabian, S. Das Sarma: Rev. Mod. Phys. **76**, 323 (2004)
2. M. Nielsen and I.L. Chuang: *Quantum Computation and Quantum Information* (Cambridge, Cambridge 2000)
3. P.W. Shor: in *Proceedings of the 35th Annual Symposium on the Foundations of Computer Science*, ed. by S. Goldwasser (IEEE Computer Society, Los Alamitos, 1994), p. 124
4. P.W. Shor: Phys. Rev A **52**, 2493 (1995); A.M. Steane: Phys. Rev. Lett. **77**, 793 (1996)
5. J.I. Cirac, P. Zoller: Phys. Rev. Lett. **74**, 4091 (1995)
6. C. Monroe, D.M. Meekhof, B.E. King, W.M. Itano, D.J. Wineland: Phys. Rev. Lett. **75**, 4714 (1995)
7. T. Sleator, H. Weinfurter: Phys. Rev. Lett. **74**, 4087 (1995)
8. Q.A. Turchette, C.J. Hood, W. Lange, H. Marbuchi, H.J. Kimble: Phys. Rev. Lett. **75**, 4710 (1995)
9. D.G. Cory, A.F. Fahmy, T.F. Havel: Proc. Natl. Acad. Sci. U.S.A. **94**, 1634 (1997); N.A. Gershenfeld, I.L. Chuang: Science **275**, 350 (1997)
10. D. Loss, D.P. DiVincenzo: Phys. Rev. A **57**, 120 (1998)
11. A. Imamoglu, D.D. Awschalom, G. Burkard, D.P. DiVincenzo, D. Loss, M. Sherwin, A. Small: Phys. Rev. Lett. **83**, 4204 (1999)
12. B.E. Kane: Nature **393**, 133 (1998); Fortschr. Phys. **48**, 1023 (2000)
13. V. Privman, I.D. Vagner, G. Kventsel: Phys. Lett. A **239**, 141 (1998)
14. R. Vrijen, E. Yablonovitch, K. Wang, H.W. Jiang, A. Balandin, V. Roychowdhury, T. Mor, D.P. DiVincenzo: Phys. Rev. A **62**, 012306 (2000)
15. A. Shnirman, G. Schön, Z. Hermon: Phys. Rev. Lett. **79**, 2371 (1997); D.V. Averin: Solid State Comm. **105**, 659 (1998)
16. D. Leibfried, B. DeMarco, V. Meyer, D. Lucas, M. Barrett, J. Britton, W.M. Itano, B. Jelenkovic, C. Langer, T. Rosenband, D.J. Wineland: Nature **422**, 412 (2003)
17. L.M.K. Vandersypen, M. Steffen, G. Breyta, C.S. Yannoni, M.H. Sherwood, I.L. Chuang: Nature **414**, 883 (2001)
18. A. Barenco, D. Deutsch, A. Ekert: Phys. Rev. Lett. **74**, 4083 (1995)
19. D.A.B. Miller, D.S. Chemla, S. Schmitt-Rink: Phys. Rev. B **33**, 6976 (1986)
20. J.A. Brum, P. Hawrylak: Superlattices and Microstructures **22**, 431 (1997)
21. G.D. Sanders, K.W. Kim, W.C. Holton: Phy. Rev. B **61**, 7526 (2000)
22. T. Tanamoto: Physica B **272**, 45 (1999); Phys. Rev. A **61**, 022305 (2000)
23. L. Fedichkin, M. Yanchenko, K.A. Valiev: Nanotechnology **11**, 387 (2000)
24. M.D. Lukin, M. Fleischhauer, R. Cote, L.M. Duan, D. Jaksch, J.I. Cirac, P. Zoller: Phys. Rev. Lett. **87**, 037901 (2001)
25. M.S. Sherwin, A. Imamoglu, T. Montroy: Phys. Rev. A **60**, 3508 (1999)
26. V. Bouchiat, D. Vion, P. Joyez, D. Esteve, M.H. Devoret: Phys. Scripta T **76**, 165 (1998)
27. Y. Nakamura, Yu.A. Pashkin, J.S. Tsai: Nature **398**, 786 (1999)
28. D. Vion, A. Aassime, A. Cottet, P. Joyez, H. Pothier, C. Urbina, D. Esteve, M.H. Devoret: Science **296**, 886 (2002)

29. T. Yamamoto, Yu.A. Pashkin, O. Astafiev, Y. Nakamura, J.S. Tsai: Nature **425**, 941 (2003)
30. T. Hayashi, T. Fujisawa, H.D. Cheong, Y.H. Jeong, Y. Hirayama: Phys. Rev. Lett. **91**, 226804 (2003)
31. S.R.E. Yang, J. Schliemann, A.H. MacDonald: Phys. Rev. B **66**, 153302 (2002)
32. V. Scarola, K. Park, S. Das Sarma: Phys. Rev. Lett. **91**, 167903 (2003)
33. P. Zanardi, F. Rossi: Phys. Rev. Lett. **81**, 4752 (1998)
34. E. Biolatti, R.C. Iotti, P. Zanardi, F. Rossi: Phys. Rev. Lett. **85**, 5647 (2000)
35. P.C. Chen, C. Piermarocchi, L.J. Sham: Phys. Rev. Lett. **87**, 067401 (2001)
36. F. Troiani, U. Hohenester, E. Molinari: Phys. Stat. Sol. B **224**, 849 (2001)
37. K.B. Brown, D.A. Lidar, K.B. Whaley: Phys. Rev. A **65**, 012307 (2002)
38. A.P. Heberle, J.J. Baumberg, K. Köhler: Phys. Rev. Lett. **75**, 2598 (1995)
39. N.H. Bonadeo, J. Erland, D. Gammon, D. Park, D.S. Katzer, D.G. Steel: Science **282**, 1473 (1998)
40. M. Bayer, P. Hawrylak, K. Hinzer, S. Fafard, M. Korkusinski, Z.R. Wasilewski, O. Stern, A. Forchel: Science **291**, 451 (2001)
41. B.E. Cole, J.B. Williams, B.T. King, M.S. Sherwin, C.R. Stanley: Nature **410**, 60 (2001)
42. G. Chen, T. H. Stievater, E.T. Batteh, X. Li, D.G. Steel, D. Gammon, D.S. Katzer, D. Park, L.J. Sham: Phys. Rev. Lett. **88**, 117901 (2002)
43. X. Li, Y.W. Wu, D.G. Steel, D. Gammon, T.H. Stievater, D.S. Katzer, D. Park, C. Piermarocchi, L.J. Sham: Science **301**, 809 (2003)
44. S. Das Sarma, J. Fabian, X. Hu, I. Žutić: Solid State Commun. **119**, 207 (2001); X. Hu, S. Das Sarma: Phys. Stat. Sol. (b) **238**, 260 (2003)
45. D.P. DiVincenzo: Science **270**, 255 (1995)
46. D. Mozyrsky, V. Privman, M. L. Glasser: Phys. Rev. Lett. **86**, 5112 (2001)
47. C. Piermarocchi, P.C. Chen, L.J. Sham, D.G. Steel: Phys. Rev. Lett. **89**, 167402 (2002)
48. T. Calarco, A. Datta, P. Fedichev, E. Pazy, P. Zoller: Phys. Rev. A **68**, 012310 (2003)
49. *Optical Orientation*, Modern Problems in Condensed Matter Science, Vol. 8, ed by F. Meier and B.P. Zakharchenya (North Holland, Amsterdam, 1984).
50. J.M. Kikkawa, D.D. Awschalom: Phys. Rev. Lett. **80**, 4313 (1998); Phys. Today **52** (6), 33 (1999)
51. D. Gammon, Al.L. Efros, T.A. Kennedy, M. Rosen, D.S. Katzer, D. Park, S.W. Brown, V.L. Korenev, I.A. Merkulov: Phys. Rev. Lett. **86**, 5176 (2001)
52. Y. Kato, R.C. Myers, A.C. Gossard, D.D. Awschalom: Nature (London) **427**, 50 (2004)
53. D.P. DiVincenzo: J. Appl. Phys. **81**, 4602 (1997)
54. D.P. DiVincenzo: J. Appl. Phys. **85**, 4785 (1999).
55. G. Burkard, D. Loss, D.P. DiVincenzo: Phys. Rev. B **59**, 2070 (1999)
56. X. Hu, S. Das Sarma: Phys. Rev. A **61**, 062301 (2000)
57. M. Friesen, P. Rugheimer, D.E. Savage, M.G. Lagally, D.W. van der Weide, R. Joynt, M.A. Eriksson: Phys. Rev. B **67**, 121301 (2003)
58. B.E. Kane, N.S. McAlpine, A.S. Dzurak, R.G. Clark, G.J. Milburn, H.B. Sun, H. Wiseman: Phys. Rev. B **61**, 2961 (2000)
59. X. Hu, R. de Sousa, S. Das Sarma: cond-mat/0108339. The Proceedings of the 7th International Symposium on Foundations of Quantum Mechanics in the Light of New Technology (Tokyo, Japan, 2001)

60. C.P. Slichter: *Principles of Magnetic Resonance*, 3rd ed (Springer, Berlin 1990)
61. A. Abragam: Principles of Nuclear Magnetism (Oxford, Oxford 1961)
62. P. Meystre and M. Sargent III: *Elements of Quantum Optics*, 2nd ed (Springer-Verlag, Berlin 1991)
63. M. Seck, M. Potemski, P. Wyder: Phys. Rev. B **56**, 7422 (1997); H.W. Jiang, E. Yablonovitch: Phys. Rev. B **64**, R041307 (2001). See also, D. Stein, K. von Klitzing, G. Weimann: Phys. Rev. Lett. **51**, 130 (1983) for an early ESR measurement in GaAs heterostructures
64. M. Thorwart, P. Hänggi: Phys. Rev. A **65**, 012309 (2001)
65. M.J. Storcz, F.K. Wilhelm: Phys. Rev. A **67**, 042319 (2003)
66. I.A. Grigorenko and D.V. Khveshchenko: Phys. Rev. Lett. **94**, 040506 (2005)
67. J.Q. You, X. Hu, F. Nori: Phys. Rev. B **72**, 144529 (2005)
68. A.C. Johnson, J.R. Petta, J.M. Taylor, A. Yacoby, M.D. Lukin, C.M. Marcus, M.P. Hanson, A.C. Gossard: Nature (London) **435**, 925 (2005)
69. J. Fabian, S. Das Sarma: J. Vac. Sci. Technol. B **17**, 1708 (1999)
70. L.A. Wu, D.A. Lidar: Phys. Rev. A **65**, 042318 (2002)
71. L.A. Wu, D.A. Lidar: Phys. Rev. A **66**, 062314 (2002)
72. A.V. Khaetskii, Yu.V. Nazarov: Phys. Rev. B **61**, 12639 (2000); *ibid.* **64**, 125316 (2001)
73. T. Fujisawa, Y. Tokura, Y. Hirayama: Phys. Rev. B **63**, 081304 (2001)
74. T. Fujisawa, D.G. Austing, Y. Tokura, Y. Hirayama, S. Tarucha: Nature **419**, 278 (2002)
75. R. Hanson, B. Witkamp, L.M.K. Vandersypen, L.H. Willems van Beveren, J.M. Elzerman, L.P. Kouwenhoven: Phys. Rev. Lett. **91**, 196802 (2003)
76. H.-A. Engel, D. Loss: Phys. Rev. Lett. **86**, 4648 (2001)
77. D. Paget, G. Lampel, B. Sapoval, V.I. Safarov: Phys. Rev. B **15**, 5780 (1977)
78. S.I. Erlingsson, Y.V. Nazarov, V.I. Fal'ko: Phys. Rev. B **64**, 195306 (2001)
79. R. de Sousa, S. Das Sarma: Phys. Rev. B **67**, 033301 (2003); *ibid.* **68**, 115322 (2003)
80. G. Bastard: *Wave mechanics applied to semiconductor heterostructures* (Halsted, New York 1988)
81. J. Schliemann, D. Loss, A.H. MacDonald: Phys. Rev. B **63**, 085311 (2001)
82. X. Hu, S. Das Sarma: Phys. Rev. A **66**, 012312 (2002)
83. K.V. Kavokin: Phys. Rev. B **64**, 075305 (2001)
84. N.E. Bonesteel, D. Stepanenko, D.P. DiVincenzo: Phys. Rev. Lett. **87**, 207901 (2001); G. Burkard, D. Loss: *ibid* **88**, 047903 (2002); D. Stepanenko, N.E. Bonesteel, D.P. DiVincenzo, G. Burkard, D. Loss: Phys. Rev. B **68**, 115306 (2003)
85. X. Hu, R. de Sousa, S. Das Sarma: Phys. Rev. Lett. **86**, 918 (2001)
86. X. Hu, S. Das Sarma: Phys. Rev. A **68**, 052310 (2003)
87. J.Levy: Phys. Rev. Lett. **89**, 147902 (2002)
88. D.P. DiVincenzo, D. Bacon, J. Kempe, G. Burkard, K.B. Whaley: Nature (London) **408**, 339 (2000)
89. D.A. Lidar, I.L. Chuang, K.B. Whaley: Phys. Rev. Lett. **81**, 2594 (1998)
90. X. Hu, S. Das Sarma: Phys. Rev. A **64**, 042312 (2001)
91. S. Vorojtsov, E.R. Mucciolo, H.U. Baranger: Phys. Rev. B **69**, 115329 (2004)
92. M. Friesen, R. Joynt, M.A. Eriksson: Appl. Phys. Lett. **81**, 4619 (2002)
93. W. Kohn, in *Solid State Physics* vol 5, page 257, ed. by F. Seitz and D. Turnbull (Academic, New York 1957)

94. K. Andres, R.N. Bhatt, P. Goalwin, T.M. Rice, R.E. Walstedt: Phys. Rev. B **24**, 244 (1981)
95. B. Koiller, X. Hu, S. Das Sarma: Phys. Rev. Lett. **88**, 027903 (2002)
96. B. Koiller, X. Hu, S. Das Sarma: Phys. Rev. B **66**, 115201 (2002)
97. C.J. Wellard, L.C.L. Hollenberg, F. Parisoli, L.M. Kettle, H.-S. Goan, J.A.L. McIntosh, D.N. Jamieson: Phys. Rev. B **68**, 195209 (2003)
98. B. Koiller, R.B. Capaz, X. Hu, S. Das Sarma: Phys. Rev. B **70**, 115207 (2004)
99. T.B. Boykin, G. Klimeck, M.A. Eriksson, M. Friesen, S.N. Coppersmith, P. von Allmen, F. Oyafuso, S. Lee: Appl. Phys. Lett. **84**, 115 (2004)
100. A.A. Larionov, L.E. Fedichkin, K.A. Valiev: Nanotechnology **12**, 536 (2001)
101. A.J. Skinner, M.E. Davenport, B.E. Kane: Phys. Rev. Lett. **90**, 087901 (2003)
102. G.D.J. Smit, S. Rogge, J. Caro, T.M. Klapwijk: Phys. Rev. B **68**, 193302 (2003)
103. A.S. Martins, R.B. Capaz, B. Koiller: Phys. Rev. B **69**, 085320 (2004)
104. D. Weinmann, W. Häusler, B. Kramer: Phys. Rev. Lett. **74**, 984 (1995)
105. M. Ciorga, M. Pioro-Ladriere, P. Zawadzki, P. Hawrylak, A.S. Sachrajda: Appl. Phys. Lett. **80**, 2177 (2002)
106. K. Ono, D.G. Austing, Y. Tokura, S. Tarucha: Science **297**, 1313 (2002)
107. A.K. Hüttel, H. Qin, A.W. Holleitner, R.H. Blick, K. Neumaier, D. Weinmann, K. Eberl, J.P. Kotthaus: Europhys. Lett. **62**, 712 (2003)
108. *Single Charge Tunneling*, ed. by H. Grabert and M.H. Devoret (Plenum, New York 1992)
109. R. Schoelkopf, P. Wahlgren, A.A. Kozhevnikov, P. Delsing, D.E. Prober: Science **280**, 1238 (1998)
110. M. Friesen, C. Tahan, R. Joynt, M.A. Eriksson: Phys. Rev. Lett. **92**, 037901 (2004)
111. E. Pazy, T. Calarco, P. Zoller: IEEE Trans. Nanotechnol. **3**, 10 (2004)
112. G. Prinz: Phys. Today **45** (4), 58 (1995)
113. P. Recher, E.V. Sukhorukov, D. Loss: Phys. Rev. Lett. **85**, 1962 (2000)
114. H.-A. Engel, V.N. Golovach, D. Loss, L.M.K. Vandersypen, J.M. Elzerman, R. Hanson, L.P. Kouwenhoven: Phys. Rev. Lett. **93**, 106804 (2004)
115. J.A. Sidles, J.L. Garbini, K.J. Bruland, D. Rugar, O. Züger, S. Hoen, C.S. Yannoni: Rev. Mod. Phys. **67**, 249 (1995)
116. B. Koiller, X. Hu, H.D. Drew, S. Das Sarma: Phys. Rev. Lett. **90**, 067401 (2003)
117. L.P. Kouwenhoven, D.G. Austing, S. Tarucha: Rep. Prog. Phys. **64**, 701 (2001)
118. S. Das Sarma, J. Fabian, X. Hu, I. Žutić: cond-mat/0002256, IEEE Trans. Magn. **36**, 2821 (2000)
119. G.B. Lesovik, T. Martin, G. Blatter: Euro. Phys. J. B **24**, 287 (2001)
120. N.M. Chtchelkatchev, G. Blatter, G.B. Lesovik, T. Martin: Phys. Rev. B **66**, 161320 (2002)
121. P. Recher, E.V. Sukhorukov, D. Loss: Phys. Rev. B **63**, 165314 (2001)
122. P. Recher, D. Loss: Phys. Rev. B **65**, 165327 (2002)
123. C. Bena, S. Vishveshwara, L. Balents, M.P.A. Fisher: Phys. Rev. Lett. **89**, 037901 (2002)
124. R. Ionicioiu, P. Zanardi, F. Rossi: Phys. Rev. A **63**, 050101 (2001)
125. W.D. Oliver, F. Yamaguchi, Y. Yamamoto: Phys. Rev. Lett. **88**, 037901 (2002)
126. D.S. Saraga, D. Loss: Phys. Rev. Lett. **90**, 166803 (2003)
127. X. Hu, S. Das Sarma: Phys. Rev. B **69**, 115312 (2004); SPIE Proceedings **5472**, 107 (2004)

128. P. Zhang, Q.K. Xue, X.G. Zhao, X.C. Xie: Phys. Rev. A **69**, 042307 (2004)
129. G. Burkard, D. Loss, E. Sukhorukov: Phys. Rev. B **61**, R16303 (2000)
130. P. Samuelsson, E. V. Sukhorukov, M. Büttiker: Phys. Rev. B **70**, 115330 (2004)
131. S. Tarucha, D.G. Austing, T. Honda, R.J. van der Hage, L.P. Kouwenhoven: Phys. Rev. Lett. **77**, 3613 (1996)
132. T. Hatano, M. Stopa, T. Yamaguchi, T. Ota, K. Yamada, S. Tarucha: Phys. Rev. Lett. **93**, 066806 (2004)
133. M. Ciorga, A.S. Sachrajda, P. Hawrylak, C. Gould, P. Zawadzki, S. Jullian, Y. Feng, Z. Wasilewski: Phys. Rev. B **61**, R16315 (2000)
134. M. Pioro-Ladriere, M. Ciorga, J. Lapointe, P. Zawadzki, M. Korkusiski, P. Hawrylak, A.S. Sachrajda: Phys. Rev. Lett. **91**, 026803 (2003)
135. J.M. Elzerman, R. Hanson, J.S. Greidanus, L.H.W. van Beveren, S. De Franceschi, L.M.K. Vandersypen, S. Tarucha, L.P. Kouwenhoven: Phys. Rev. B **67**, 161308 (2003)
136. J.R. Petta, A.C. Johnson, C.M. Marcus, M.P. Hanson, A.C. Gossard: Phys. Rev. Lett. **93**, 186802 (2004)
137. J.M. Elzerman, R. Hanson, L.H.W. van Beveren, L.M.K. Vandersypen, L.P. Kouwenhoven: Appl. Phys. Lett. **84**, 4617 (2004)
138. J. M. Elzerman, R. Hanson, L.H.W. van Beveren, B. Witkamp, L.M.K. Vandersypen, L.P. Kouwenhoven: Nature **430**, 431 (2004)
139. R. Hanson, L.H. Willems van Beveren, I.T. Vink, J.M. Elzerman, W.J.M. Naber, F.H.L. Koppens, L.P. Kouwenhoven, L.M.K. Vandersypen: Phys. Rev. Lett. **94**, 196802 (2005)
140. I.A. Merkulov, Al.L. Efros, M. Rosen: Phys. Rev. B **65**, 205309 (2002)
141. F. Jelezko, I. Popa, A. Gruber, C. Tietz, J. Wrachtrup, A. Nizovtsev, S. Kilin: Appl. Phys. Lett. **81**, 2160 (2002)
142. H.J. Mamin, R. Budakian, B.W. Chui, D. Rugar: Phys. Rev. Lett. **91**, 207604 (2003)
    D. Rugar, R. Budakian, H.J. Mamin, B.W. Chui: Nature **430**, 329 (2004)
143. R. Budakian, H.J. Mamin, D. Rugar: Phys. Rev. Lett. **92**, 037205 (2004)
144. D. Gammon, S.W. Brown, E.S. Snow, T.A. Kennedy, D.S. Katzer, D. Park: Science **277**, 85 (1997)
145. R.K. Kawakami, Y. Kato, M. Hanson, I. Malajovich, J.M. Stephens, E. Johnston-Halperin, G. Salis, A.C. Gossard, D.D. Awschalom: Science **294**, 131 (2001)
146. K. Ono, S. Tarucha: Phys. Rev. Lett. **92**, 256803 (2004)
147. J.R. Petta, A.C. Johnson, J.M. Taylor, E.A. Laird, A. Yacoby, M.D. Lukin, C.M. Marcus, M.P. Hanson, A.C. Gossard: Science **309**, 2180 (2005)
148. R.G. Clark, R. Brenner, T.M. Buehler, V. Chan, N.J. Curson, A.S. Dzurak, E. Gauja, H.S. Goan, A.D. Greentree, T. Hallam, A.R. Hamilton, L.C.L. Hollenberg, D.N. Jamieson, J.C. McCallum, G.J. Milburn, J.L. O'Brien, L. Oberbeck, C.I. Pakes, S.D. Prawer, D.J. Reilly, F.J. Ruess, S.R. Schofield, M.Y. Simmons, F.E. Stanley, R.P. Starrett, C. Wellard, C. Yang: Phil. Trans. Roy. Soc. A **361**, 1451 (2003)
149. *Proceedings of the 1st International Conference on Experimental Implementation of Quantum Computation*, ed. by R.G. Clark (Rinton, Princeton 2001)
150. M. Altarelli, W.L. Hsu, Phys. Rev. Lett. **43**, 1346 (1979)
151. T. Schenkel, A. Persaud, S.J. Park, J. Meijer, J.R. Kingsley, J.W. McDonald, J.P. Holder, J. Bokor, D.H. Schneider: J. Vac. Sci. Technol B **20**, 2819 (2002)

152. C.Y. Yang, D.N. Jamieson, C. Pakes, S. Prawer, A. Dzurak, F. Stanley, P. Spizziri, L. Macks, E. Gauja, R.G. Clark: Jpn. J. Appl. Phys. (part 1) **42** (6B), 4124 (2003)
153. J.L. O'Brien, S.R. Schofield, M.Y. Simmons, R.G. Clark, A.S. Dzurak, N.J. Curson, B.E. Kane, N.S. McAlpine, M.E. Hawley, G.W. Brown: Phys. Rev. B **64**, 161401 (2001)
154. S.R. Schofield, N.J. Curson, M.Y. Simmons, F.J. Rueβ, T. Hallam, L. Oberbeck, R.G. Clark: Phys. Rev. Lett. **91**, 136104 (2003)
155. V.W. Scarola, S. Das Sarma: Phys. Rev. A **71**, 032340 (2005)
156. M. Stopa: unpublished

Microscopic Theory of Coherent Semiconductor Optics

T. Meier and S.W. Koch

Department of Physics and Material Sciences Center, Philipps University,
Renthof 5, 35032 Marburg, Germany

Abstract. The derivation of a microscopic many-body theory for the nonlinear optical response of semiconductors is reviewed. At the Hartree–Fock level, the semiconductor Bloch equations include many-body effects via band gap and field renormalization. These equations are sufficient to describe excitonic resonances as they appear already in the linear absorption spectra. An adequate description of nonlinear optical effects in semiconductors beyond the Hartree–Fock level includes Coulomb interaction induced carrier correlations. Different schemes have been developed to treat such correlation effects. As two examples, the second-order Born approximation and the dynamics-controlled truncation scheme are introduced and analyzed. In addition to the derivation of the equations of motion, a few examples are presented which highlight important signatures of many-body correlations in the optical response of semiconductors.

1 Introduction

This article reviews a microscopic many-body theory which is capable of describing the nonlinear optical response of semiconductors. It is a brief version of topics which have been discussed in [1, 2, 3, 4] in more detail. Here, we concentrate on describing the main ingredients of the theory. Additionally, several examples which highlight some important consequences of the introduced theoretical approaches are presented. Many more examples and comparisons between calculated and measured signals can be found in the references given above and in the original publications.

For a comprehensive analysis of the optical properties of semiconductors it is required to describe the light field and material excitation dynamics as well as their interactions [1]. On the semiclassical level of the theory, the light field is treated classically at the level of Maxwell's equations, whereas the material excitations are computed quantum mechanically.

The electromagnetic field and the optical polarization have to be calculated self-consistently. In dipole approximation, the light-matter coupling is given by the scalar product of the electric field and the optical polarization of the medium. The induced optical polarization then appears as a source term for the electromagnetic field in Maxwell's wave equations. Hence, the polarization generates fields which depend on the state of the material system and

T. Meier and S.W. Koch: *Microscopic Theory of Coherent Semiconductor Optics*,
Lect. Notes Phys. **689**, 115–152 (2006)
www.springerlink.com

therefore yield information about the medium excitations. Since the light-matter interaction is mediated by the optical polarization, this polarization is the key quantity in the description of the material excitations.

Here, we focus on situations where the semiconductors are excited with optical fields which have frequencies in the vicinity of their fundamental band gap. For such interband excitation conditions, the optical polarization of a semiconductor can be described as superposition of coherences between valence- and conduction-band states. The linear optical absorption spectrum of an unexcited semiconductor is determined by the sum of all interband transitions between the completely filled valence bands and the empty conduction bands. Already in this regime, one finds characteristic signatures which are due to the Coulomb interaction among the photoexcited carriers [1, 5]. In particular, the Coulomb attraction between optically excited electrons and holes leads to strongly absorbing resonances that appear spectrally below the band gap. These lines are due to bound electron-hole-pair states, which are denoted as excitons. Thus, the linear absorption of an unexcited semiconductor already shows that the Coulomb interaction has to be treated properly if one wants to achieve a meaningful theory for the optical response.

If the semiconductor is not in its ground state or if the exciting light fields are not in the low-intensity limit, one has to go beyond the linear optical regime described above. In this case, additional material quantities besides the interband polarization need to be considered for a consistent analysis of the nonlinear optical response. Since the Coulomb interaction introduces a many-body problem, it is usually not possible to perform a fully exact treatment of the nonlinear response. However, during the last decades, a number of approximation schemes have been developed and applied successfully to different excitation regimes, see, e.g., [3, 6, 7, 8, 9, 10, 11, 12, 13, 14, 15, 16, 17].

The simplest approximation that can be used to describe a many-body system is to model it as an effective single-particle system with renormalized potentials. This is effectively done when the optical response of semiconductors is analyzed with the time-dependent Hartree–Fock theory. On this level, the material dynamics is completely determined by the diagonal and off-diagonal elements of the reduced single-particle density matrix [1]. While the off-diagonal terms of the density matrix are given by interband transitions, which determine the optical polarization, the diagonal terms are the carrier (electron and hole) occupation probabilities in the different bands. The temporal dynamics of the optically induced excitations is obtained from solutions of the coupled equations of motion for the density-matrix elements, which are the Hartree–Fock semiconductor Bloch equations. These equations include Coulomb effects in first order via renormalizations of the band gap and the optical field [1, 7, 18] and are well suited to analyze a number of important effects, in particular, in the coherent nonlinear optical response of semiconductors.

All properties of the many-body system that cannot be described on the basis of the Hartree–Fock approximation are denoted as *correlation effects* [19]. Similarly, all aspects of the nonlinear optical semiconductor response that cannot be described on the basis of the Hartree–Fock semiconductor Bloch equations are caused by the many-body correlations. If the analysis goes beyond the linear optical properties of unexcited semiconductors, such correlation effects are always present and lead, in certain situations, to significant experimentally observable consequences. Different approaches have been developed to deal with these many-body correlations. An approach which is particularly useful in the regime of strong fields is the second-order Born approximation [7, 11, 13, 14, 15, 17, 20]. Density-dependent absorption spectra showing exciton saturation and broadening as well as gain spectra of semiconductor lasers are examples for which the results of the second-order Born approximation are in good agreement with experiment [2, 4, 21]. Another approach, which is based on an expansion of the nonlinear optical response in powers of the optical field [22], is the so-called dynamics-controlled truncation scheme [3, 4, 9, 10, 12, 16, 23]. This scheme is often used together with the assumption of a fully coherent system, which strongly reduces the number of dynamic variables, since in this case only single- and multi-exciton coherences have to be considered.

At the fully quantum mechanical level, one needs to take into account correlations arising from the light-matter interaction in addition to the many-body Coulomb correlations discussed above. The combination of these correlation effects can be described by coupled semiconductor Bloch and luminescence equations. In these equations, not only the material excitations but also the light field are treated quantum mechanically. Since we concentrate in the following on the analysis of semiclassical aspects of semiconductor optics, we do not discuss this approach here. Instead, we refer the interested reader to the literature, e.g., [24, 25, 26, 27].

Quite recently, it has been shown that the second-order Born approximation and the dynamics-controlled truncation scheme, can be obtained as limiting cases of a unified theoretical approach. This unified theory treats light-matter and Coulomb many-body correlations on the basis of a cluster expansion [28]. If this analysis is restricted to classical light fields, the light-matter correlations vanish. Considering furthermore the interaction with the optical field only up to a finite order and assuming a fully coherent system, the dynamics-controlled truncation scheme can be obtained from the cluster expansion. To obtain the second-order Born approximation, one has to approximate the four-particle correlations which appear in the cluster expansion, e.g., biexciton transitions, exciton occupations, and fourth-order intraband correlations, by factorizing their sources into products of polarizations and electron and hole occupations. Coupled equations which contain the dynamics-controlled truncation scheme and the second-order Born approximation as limiting cases have also been obtained using an approach which is based on non-equilibrium Green's functions, see [29, 30].

In the following, we introduce the fundamental concepts which are required for the description of the optical semiconductor response in the semiclassical approach. In Sect. 2, it is shown that within the semiclassical theory the optical polarization of the semiconductor appears as an inhomogeneity in the wave equation of the electric field. As discussed above, due to the Coulomb interaction among the carriers, one has to solve a many-body problem to describe semiconductor optics. Sect. 3 reviews the derivation of the semiconductor Bloch equations on the Hartree–Fock level. These equations are then used in Sect. 3.1 for the analysis of the excitonic absorption spectra of semiconductor nanostructures of different dimensionality.

The theoretical treatment of many-body Coulomb correlations is reviewed in Sect. 4. First, the second-order Born approximation is discussed in Sect. 4.1. This approach is capable of describing the nonlinear optical response semiconductor nanostructures in various situations. As examples, we present numerical results on optical absorption spectra and the temporal dynamics of the polarization under different excitation conditions in Sect. 4.2. In Sect. 4.3, the dynamics-controlled truncation scheme is introduced. The relevant equations of motion up to third-order in the field are derived for a coherent system, i.e., the coherent $\chi^{(3)}$-limit. In this approach, biexcitonic four-particle correlations are treated explicitly. Some experimentally observable consequences arising from the dynamics of these four-particle correlations are presented in Sect. 4.4. Section 4.5 outlines how one can extend the dynamics-controlled truncation scheme to include terms up to fifth-order in the interaction with the field, i.e., the coherent $\chi^{(5)}$-limit. A few signatures of fifth-order biexcitonic correlations are presented in Sect. 4.6.

The final Sect. 5 contains conclusions and a brief outlook on future challenges in the area of correlation effects in semiconductor optics.

2 Semiclassical Theory

For the computation of the optical response of a material system, the solution of Maxwell's equations is required. For the simple example of a one-dimensional propagation in z direction, the wave equation for the electric field is given by

$$\left[\frac{\partial^2}{\partial z^2} + \frac{n^2(z)}{c^2} \frac{\partial^2}{\partial t^2} \right] E(z,t) = -\mu_0^{eh} \frac{\partial^2}{\partial t^2} P(z,t), \tag{1}$$

where E is the electric field, P the macroscopic optical polarization of the medium, t the time, and μ_0^{eh} is a prefactor which depends on the system of units. Both E and P are two-dimensional vectors in the x- y-plane, representing the polarization direction.

In (1), the material response has been divided into two parts. The resonant material excitations are included dynamically in terms of the optical polarization P, whereas all nonresonant contributions are included in the background refractive index n. From (1) it is clear that the polarization P is the key quantity which needs to be determined adequately when describing the material response. In the following, a few concepts are described which allow one to obtain P from microscopic theory.

3 Time-Dependent Hartree–Fock Approximation

For ordered semiconductors and semiconductor nanostructures, e.g., quantum wells, quantum wires, and superlattices, it is usually convenient to expand the macroscopic optical polarization into a Bloch basis [1]

$$\mathbf{P} = \sum_{\mathbf{k},e,h} (\boldsymbol{\mu}^{eh})^* \, p_{\mathbf{k}}^{eh} + \text{c.c.}, \tag{2}$$

where $\boldsymbol{\mu}^{eh}$ is the electron-hole interband dipole matrix element between conduction band e and valence band h. For the development of the many-body theory, we use the electron $(a_{e,\mathbf{k}}^\dagger, a_{e,\mathbf{k}})$ and hole $(b_{h,\mathbf{k}}^\dagger, b_{h,\mathbf{k}})$ creation and annihilation operators. The microscopic polarization $p_{\mathbf{k}}^{eh}$ and the carrier occupation probabilities $f_{\mathbf{k}}^{ee,hh}$ constitute the diagonal and off-diagonal elements of the reduced single-particle density matrix:

$$\begin{pmatrix} \langle a_{e,\mathbf{k}}^\dagger a_{e,\mathbf{k}} \rangle & \langle b_{h,-\mathbf{k}} a_{e,\mathbf{k}} \rangle \\ \langle a_{e,\mathbf{k}}^\dagger b_{h,-\mathbf{k}}^\dagger \rangle & \langle b_{h,-\mathbf{k}}^\dagger b_{h,-\mathbf{k}} \rangle \end{pmatrix} = \begin{pmatrix} f_{\mathbf{k}}^{ee} & p_{\mathbf{k}}^{eh} \\ (p_{\mathbf{k}}^{eh})^* & f_{\mathbf{k}}^{hh} \end{pmatrix}. \tag{3}$$

The equations of motion describing the dynamical evolution of the components of the density matrix are given by the Heisenberg equation, which for an arbitrary operator $\mathcal{A}$ reads

$$i\hbar \frac{\partial}{\partial t} \mathcal{A} = [\mathcal{A}, H]. \tag{4}$$

For the analysis of the optical properties of semiconductors, we use the standard many-body Hamiltonian [1]

$$H = H_{\text{single–particle}} + H_{\text{Coulomb}} + H_{\text{light–matter}}. \tag{5}$$

In (5),

$$H_{\text{single–particle}} = \sum_{\mathbf{k},e} \varepsilon_{\mathbf{k}}^e a_{e,\mathbf{k}}^\dagger a_{e,\mathbf{k}} + \sum_{\mathbf{k},h} \varepsilon_{\mathbf{k}}^h b_{h,-\mathbf{k}}^\dagger b_{h,-\mathbf{k}} \tag{6}$$

is the single-particle Hamiltonian, which contains the band structure (single-particle energy) of the electrons $(\varepsilon_{\mathbf{k}}^e)$ in conduction band e and holes $(\varepsilon_{\mathbf{k}}^h)$ in valence band h. H_{Coulomb} describes the Coulomb interaction between the carriers and is given by

$$H_{\text{Coulomb}} = \frac{1}{2} \sum_{\mathbf{k},\mathbf{k}',\mathbf{q}\neq 0, e, e'} V_q a^\dagger_{e,\mathbf{k}+\mathbf{q}} a^\dagger_{e',\mathbf{k}'-\mathbf{q}} a_{e',\mathbf{k}'} a_{e,\mathbf{k}}$$

$$+ \frac{1}{2} \sum_{\mathbf{k},\mathbf{k}',\mathbf{q}\neq 0, h, h'} V_q b^\dagger_{h,\mathbf{k}+\mathbf{q}} b^\dagger_{h',\mathbf{k}'-\mathbf{q}} b_{h',\mathbf{k}'} b_{h,\mathbf{k}}$$

$$- \sum_{\mathbf{k},\mathbf{k}',\mathbf{q}\neq 0, e, h} V_q a^\dagger_{e,\mathbf{k}+\mathbf{q}} b^\dagger_{h,\mathbf{k}'-\mathbf{q}} b_{h,\mathbf{k}'} a_{e,\mathbf{k}} . \tag{7}$$

Here, the first two lines denote the repulsive interactions among two electrons and among two holes, respectively, and the third line describes the attractive interactions between electrons and holes. V_q denotes the Fourier transform of the Coulomb potential. In dipole approximation, the interaction between the carrier system and the classical electromagnetic field is described by [1]

$$H_{\text{light}-\text{matter}} = -\mathbf{E}(t) \cdot \sum_{\mathbf{k},e,h} (\boldsymbol{\mu}^{eh} a^\dagger_{e,\mathbf{k}} b^\dagger_{h,-\mathbf{k}} + (\boldsymbol{\mu}^{eh})^* b_{h,-\mathbf{k}} a_{e,\mathbf{k}}), \tag{8}$$

with the electron-hole interband dipole matrix element $\boldsymbol{\mu}^{eh}$. The sum that appears in (8) yields the macroscopic optical polarization, as defined in (2). (8) shows that the light field $\mathbf{E}(t)$ either creates or destroys *pairs* of electrons and holes.

For simplicity, we neglect here optical transitions involving light-holes and energetically deeper valence bands. Thus, only the heavy-hole valence band and the lowest conduction band, both of which are twofold spin-degenerate, are taken into account. The two heavy-hole bands ($h = 1, 2$) are described by the states $|-3/2, h >$ and $|3/2, h >$, and the lowest conduction bands ($e = 1, 2$) by $|-1/2, e >$ and $|1/2, e >$, respectively [31, 32]. For light which propagates in the z-direction, i.e., perpendicular to the plane of the quantum well, we use the usual circularly polarized dipole matrix elements [31, 32]:

$$\boldsymbol{\mu}^{11} = \mu_0 \, \boldsymbol{\sigma}^+ = \frac{\mu_0}{\sqrt{2}} \begin{pmatrix} 1 \\ i \end{pmatrix},$$

$$\boldsymbol{\mu}^{12} = \boldsymbol{\mu}^{21} = 0,$$

$$\boldsymbol{\mu}^{22} = \mu_0 \, \boldsymbol{\sigma}^- = \frac{\mu_0}{\sqrt{2}} \begin{pmatrix} 1 \\ -i \end{pmatrix}, \tag{9}$$

where μ_0 is the modulus of $\boldsymbol{\mu}^{11}$ and $\boldsymbol{\mu}^{22}$. Due to these *diagonal* selection rules, i.e., $\boldsymbol{\mu}^{eh} \propto \delta_{eh}$, there are two separate subspaces of optical excitations. All the transitions are, however, coupled by the many-body Coulomb interaction, since it is independent of the band indices (spin), see (7).

By evaluating the commutators of (4), one obtains the equation of motion for the microscopic polarization

$$i\hbar\frac{\partial}{\partial t}p_{\mathbf{k}}^{eh} = -\left(\varepsilon_{\mathbf{k}}^e + \varepsilon_{\mathbf{k}}^h\right)p_{\mathbf{k}}^{eh} + \sum_{\mathbf{q}\neq 0}V_q p_{\mathbf{k}-\mathbf{q}}^{eh}$$

$$+ \left(\boldsymbol{\mu}^{eh} - \sum_{e'}f_{\mathbf{k}}^{ee'}\boldsymbol{\mu}^{e'h} - \sum_{h'}\boldsymbol{\mu}^{eh'}f_{\mathbf{k}}^{h'h}\right)\cdot\mathbf{E}$$

$$- \sum_{\mathbf{q}\neq 0,\mathbf{k}',e'}V_q\left[\left\langle a_{e,\mathbf{k}}^{\dagger}a_{e',\mathbf{k}'}^{\dagger}b_{h,\mathbf{k}+\mathbf{q}}^{\dagger}a_{e',\mathbf{k}'-\mathbf{q}}\right\rangle\right.$$

$$\left. - \left\langle a_{e,\mathbf{k}+\mathbf{q}}^{\dagger}a_{e',\mathbf{k}'}^{\dagger}b_{h,\mathbf{k}}^{\dagger}a_{e',\mathbf{k}'+\mathbf{q}}\right\rangle\right]$$

$$+ \sum_{\mathbf{q}\neq 0,\mathbf{k}',h'}V_q\left[\left\langle a_{e,\mathbf{k}+\mathbf{q}}^{\dagger}b_{h',\mathbf{k}'+\mathbf{q}}^{\dagger}b_{h,\mathbf{k}}^{\dagger}b_{h',\mathbf{k}'}\right\rangle\right.$$

$$\left. - \left\langle a_{e,\mathbf{k}}^{\dagger}b_{h',\mathbf{k}'+\mathbf{q}}^{\dagger}b_{h,\mathbf{k}-\mathbf{q}}^{\dagger}b_{h',\mathbf{k}'}\right\rangle\right]. \tag{10}$$

For a two-band model, i.e., if only a single conduction and a single valence band is considered, we have $e = e' = 1$ and $h = h' = 1$ in (10). In the more general multiband case, the summations over the respective bands, e' and h', have to be considered. Since this analysis is restricted to transitions from heavy-holes to the lowest conduction band, $p_{\mathbf{k}}^{eh}$ is nonvanishing only for $e = h = 1$ and $e = h = 2$. Thus, concerning the subband indices, $p_{\mathbf{k}}^{eh}$ is proportional to δ_{eh}, since the terms with $e \neq h$ have no sources. For similar reasons one can show that also $f_{\mathbf{k}}^{ee'}$ and $f_{\mathbf{k}}^{h'h}$ are diagonal, i.e., proportional to $\delta_{ee'}$ and $\delta_{hh'}$, respectively. Therefore, it would be possible to slightly reduce the notation by defining $f_{\mathbf{k}}^e \equiv f_{\mathbf{k}}^{ee}$ and $f_{\mathbf{k}}^h \equiv f_{\mathbf{k}}^{hh}$. For clarity, we keep, however, $f_{\mathbf{k}}^{ee}$ and $f_{\mathbf{k}}^{hh}$ in the following, and switch to the notation $f_{\mathbf{k}}^e$ and $f_{\mathbf{k}}^h$ at a later stage, when the second-order Born approximation is treated.

Equation (10) and the corresponding equations for the carrier occupation probabilities, $f_{\mathbf{k}}^{ee}$ and $f_{\mathbf{k}}^{hh}$, contain a coupling of the components of the single-particle density matrix among themselves (via $H_{\text{single-particle}}$ and $H_{\text{light-matter}}$) and contributions that couple the two-operator terms to four-operator terms (via H_{Coulomb}) [1]. The appearance of those four-operator terms is the beginning of the well known many-body hierarchy. If one computes equations of motion for $2n$ operator expectation values, they are coupled due to the many-body Coulomb interaction to $2n + 2$ operator terms, and so on. In order to close such sets of coupled equations for many-body systems, one has to use approximations, i.e., the hierarchy has to be truncated at some stage in a self-consistent fashion.

On the level of the time-dependent Hartree–Fock approximation, one uses a decoupling scheme which factorizes the four-operator terms into a product of two two-operator terms, e.g.,

$$\left\langle a_{e,\mathbf{k}}^{\dagger}a_{e,\mathbf{k}'-\mathbf{q}}^{\dagger}a_{e,\mathbf{k}-\mathbf{q}}a_{e,\mathbf{k}'}\right\rangle \simeq \left\langle a_{e,\mathbf{k}}^{\dagger}a_{e,\mathbf{k}}\right\rangle\left\langle a_{e,\mathbf{k}-\mathbf{q}}^{\dagger}a_{e,\mathbf{k}-\mathbf{q}}\right\rangle\delta_{\mathbf{k},\mathbf{k}'}, \tag{11}$$

where no other contribution appears since $\mathbf{q} \neq 0$ [1]. Applying this procedure to all terms that appear in the equation of motion for the microscopic polarization, (10), and the corresponding equations for the carrier occupations, one obtains the well-known semiconductor Bloch equations in time-dependent Hartree–Fock approximation. For a two band system, i.e., $e = e' = h = h' = 1$, these equations can be written as [1, 7]

$$\left[i\hbar\frac{\partial}{\partial t} - \epsilon_{\mathbf{k}}^e(t) - \epsilon_{\mathbf{k}}^h(t)\right] p_{\mathbf{k}}^{eh}(t) = \left[1 - f_{\mathbf{k}}^{ee}(t) - f_{\mathbf{k}}^{hh}(t)\right] \Omega_{\mathbf{k}}(t),$$

$$\frac{\partial}{\partial t} f_{\mathbf{k}}^{aa}(t) = -\frac{2}{\hbar}\mathrm{Im}\left[\Omega_{\mathbf{k}}(t)(p_{\mathbf{k}}^{eh}(t))^*\right], \tag{12}$$

where

$$\Omega_{\mathbf{k}}(t) = \boldsymbol{\mu}^{eh}\cdot\mathbf{E}(t) + \sum_{\mathbf{k'}\neq\mathbf{k}} V_{|\mathbf{k}-\mathbf{k'}|}\, p_{\mathbf{k'}}^{eh}(t) \tag{13}$$

is the renormalized field (Rabi frequency) and

$$\epsilon_{\mathbf{k}}^a(t) = \varepsilon_{\mathbf{k}}^a - \sum_{\mathbf{k'}\neq\mathbf{k}} V_{|\mathbf{k}-\mathbf{k'}|}\, f_{\mathbf{k'}}^{aa}(t) \tag{14}$$

the renormalized single-particle energy. If one ignores the Coulomb interaction ($V \equiv 0$) in (12)–(14), they become diagonal in $\mathbf{k}$. In this artificial case, the optical response is described by the superposition of uncoupled two-level systems in $\mathbf{k}$ space and the band structure simply provides an inhomogeneous broadening. For finite Coulomb interaction, the Hartree–Fock terms, i.e., exchange renormalizations, lead to a coupling among all $\mathbf{k}$-states of the semiconductor material and introduce nonlinearities in the Hartree–Fock semiconductor Bloch equations. It should be noted that the renormalization contributions are by no means small corrections. They contain, in particular, the leading-order Coulomb effect which gives rise to excitonic resonances, as shown in Sect. 3.1.

The optical nonlinearities contained in the Hartree–Fock semiconductor Bloch equations, (12), are due to phase-space filling, which is represented by the terms proportional to the occupation probabilities times the field, i.e., Ef, as well as energy and field renormalizations. The phase-space filling is a consequence of the Fermionic nature of the electrons and holes (Pauli blocking), whereas the renormalizations arise due to anharmonic terms in the Hamiltonian (terms containing more than two operators, i.e., H_{Coulomb}). It should be noted, that on the Hartree–Fock level the semiconductor Bloch equations contain neither dephasing of the polarization nor screening of the interaction potential or relaxation of the carrier distributions. For the description of these effects, it is required to go beyond the Hartree–Fock approximation and to treat many-body correlations, as is outlined below in Sect. 4.

3.1 Excitonic Linear Absorption Spectra in Different Dimensions

Starting from an unexcited semiconductor, i.e., neither polarization nor electron and hole populations are present before the system is excited, one can linearize the polarization equation in the interaction with the external field. For parabolic bands (effective mass approximation), the linear optical polarization is determined by

$$\left[i\hbar \frac{\partial}{\partial t} - \epsilon_g - \frac{\hbar^2 k^2}{2m_r} \right] p_{\mathbf{k}}^{eh}(t) = \boldsymbol{\mu}^{eh} \cdot \mathbf{E}(t) + \sum_{\mathbf{k}'} V_{|\mathbf{k}-\mathbf{k}'|}\, p_{\mathbf{k}'}^{eh}(t) , \qquad (15)$$

where ϵ_g is the band gap energy and $m_r = (1/m_e + 1/m_h)^{-1}$ is the reduced mass, which is the inverse of the sum over the inverse effective masses of the valence and conductions bands, m_h and m_e, respectively. By Fourier transforming (15) to real space, one obtains

$$\left[i\hbar \frac{\partial}{\partial t} - \epsilon_g + \frac{\hbar^2 \nabla_{\mathbf{r}}^2}{2m_r} - V(r) \right] p^{eh}(\mathbf{r}, t) = -\boldsymbol{\mu}^{eh} \cdot \mathbf{E}(t)\, \delta(\mathbf{r}) . \qquad (16)$$

The homogeneous part of (16), i.e., its left hand side, is the *Wannier equation*. It is mathematically identical to the Schrödinger equation for the relative motion of proton and electron in a hydrogen atom. Hence, the solutions of the Wannier equation contain a discrete Rydberg series of bound eigenstates (energies smaller than ϵ_g) and unbound solutions (energies bigger than ϵ_g). The bound states are referred to as Wannier excitons and the unbound states represent the Coulomb interacting continuum interband excitations. By solving the inhomogeneous (16), one obtains the linear electron-hole pair susceptibility $\chi^{(1)}(\omega) = P(\omega)/E(\omega)$. The imaginary part of the susceptibility, $Im[\chi^{(1)}(\omega)]$, is proportional to the linear absorption $\alpha^{(1)}(\omega)$ as described by the Elliott formula [5].

The linear optical absorption spectra for three-, two-, and one-dimensional direct-gap semiconductors, corresponding to bulk, quantum well, and quantum wire structures, respectively, are shown in Fig. 1 in the spectral vicinity of the band gap. The shape of the dashed lines is relatively simple to understand. As already noted above, without Coulomb interaction, the microscopic polarizations $p_{\mathbf{k}}^{eh}$ for different $\mathbf{k}$ do not interact, see (12) and (13). Therefore, in this case, the interband transitions are described by noninteracting two-level systems and the dispersion of the valence and conduction bands just introduces an inhomogeneous broadening. As a consequence, the absorption vanishes for frequencies that are smaller than the band gap ϵ_g. Above the gap, the absorption profile is proportional to the joint density of states for the valence to conduction band transitions, i.e., the absorption is proportional to $\Theta(\hbar\omega - \epsilon_g)(\hbar\omega - \epsilon_g)^{(d-2)/2}$ where d is the effective dimensionality of the carrier dynamics [1].

Including the Coulomb interaction modifies the spectra significantly. In any dimension, a hydrogenic series of strongly absorbing exciton states appears as discrete lines *below* the band gap, see Fig. 1. As a consequence of the optical selection rules, i.e., the coupling to the light field is proportional to $\delta(\mathbf{r})$ in (16), one typically observes in direct semiconductors only *s*-states

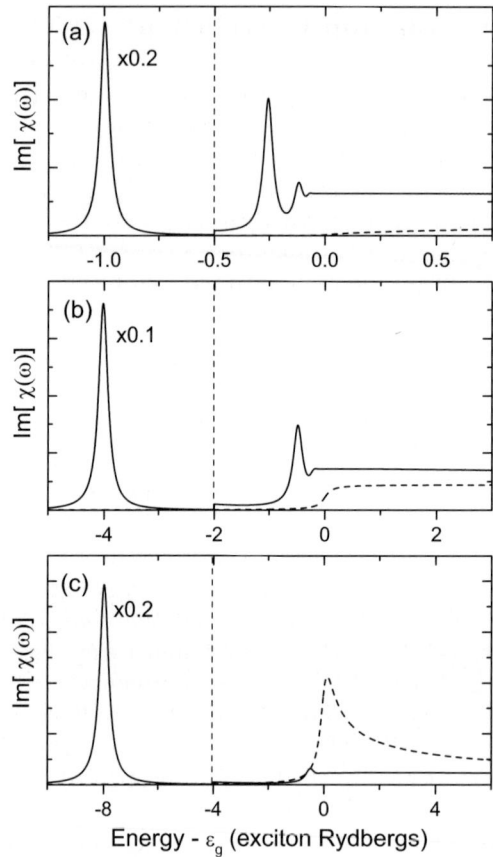

Fig. 1. Imaginary part of the linear susceptibility, $Im[\chi^{(1)}(\omega)]$, which is proportional to the linear optical absorption, $\alpha^{(1)}(\omega)$, in the vicinity of the band gap for a (**a**) three-, (**b**) two-, and (**c**) one-dimensional direct semiconductor, corresponding to bulk, quantum well, and quantum wire structures, respectively. The *dashed lines* are calculated neglecting excitonic effects (vanishing Coulomb interaction), whereas the *solid lines* represent the results of the full calculations. For each case a homogeneous broadening corresponding to 1/20th of the respective exciton binding energy has been used. For better visibility of the absorption close to the band edge, the peak due to the 1 s exciton (*left* of *dashed lines*) has been reduced by a factor as indicated. Taken from [33]

of the hydrogenic series. In addition, the Coulomb interaction also modifies the absorption at and above ϵ_g. Instead of the joint density of states seen in the noninteracting case, the absorption due to transitions to unbound, but interacting electron-hole pairs varies only weakly as function of energy. More details and analytical evaluations for bulk systems and semiconductor nanostructures can be found, e.g., in [1].

4 Many-Body Coulomb Correlations

In this section, two theoretical approaches that are capable of describing many-body correlation effects are reviewed. We first focus on the so-called second-order Born approximation. Then the dynamics-controlled truncation scheme in the coherent third-order $(\chi^{(3)})$ and fifth-order $(\chi^{(5)})$ limits is described. For each case, besides the derivation of the equations, a few examples for experimentally observable consequences of the respective many-body correlation contributions are presented and discussed.

4.1 Second-Order Born Approximation

As shown in Sect. 3, on the level of the time-dependent Hartree–Fock approximation the many-body Coulomb interaction leads to renormalizations of the single-particle energies and of the Rabi frequency, i.e., each carrier evolves in an effective medium which is provided by the other carriers. These renormalizations, which are of first-order in the Coulomb interaction, introduce a coupling of all states in $\mathbf{k}$-space and lead to a strong dependence of the nonlinear optical response on the excitation conditions. However, the Hartree–Fock semiconductor Bloch equations are coherent in the sense that the first-order Coulomb terms describe neither the dephasing of the microscopic polarizations nor the relaxation of nonequilibrium carrier populations.

If one wants to analyze these important effects on a microscopic basis it is required to treat many-body correlations which are beyond the effective-medium Hartree–Fock level. In what follows it is shown that dephasing and relaxation processes can be analyzed if one treats the terms which are of second order in the many-body Coulomb interaction, i.e., the second-order Born approximation. The resulting equations have the general structure of quantum Boltzmann-type equations and describe, e.g., carrier-carrier scattering processes. Furthermore, they also include scattering among the microscopic polarizations, which are complex quantities, and coupling between the polarizations and populations. Unlike in the Hartree–Fock treatment, on this level of theory photoexcitations at, e.g., $\mathbf{k}$ and $\mathbf{k}'$ are explicitly coupled and transferred to a certain extend into excitations at $\mathbf{k} + \mathbf{q}$ and $\mathbf{k}' - \mathbf{q}$.

The simplest way to treat relaxation and dephasing is to phenomenologically add the terms

$$\frac{\partial}{\partial t} p_{\mathbf{k}}^{eh}(t)|_{corr} = -\frac{p_{\mathbf{k}}^{eh}(t)}{T_2},$$
$$\frac{\partial}{\partial t} f_{\mathbf{k}}^{aa}(t)|_{corr} = -\frac{f_{\mathbf{k}}^{aa}(t) - F_{\mathbf{k}}^{aa}}{T_{rel}}, \tag{17}$$

with $aa = ee$ and $aa = hh$ to the Hartree–Fock semiconductor Bloch equations, (12). These terms lead to an exponential decay of the polarization with the decay (dephasing) time T_2 and a thermalization of the actual carrier distribution towards the quasi-equilibrium distribution $F_{\mathbf{k}}^{aa}$ with the relaxation time T_{rel}. Even though the assumption of such simple decay and relaxation processes are sometimes compatible with experimental information, the systematic derivation and computation of relaxation and dephasing times is not straightforward. As is shown below, in the microscopic approach, the decay and relaxation terms are $\mathbf{k}$-dependent and depend on the excitations that are present in the systems, i.e., they are complicated functionals of $p_{\mathbf{k}}^{eh}$ and $f_{\mathbf{k}}^{aa}$.

A microscopically consistent way to obtain approximations beyond the Hartree–Fock level is to derive equations of motion for the four-particle correlation functions and to use systematic factorization procedures at a higher level. The derivation can be done, e.g., by using Green's function techniques, projection operator, or equation of motion methods [7, 11, 13, 14, 34]. The second-order Born approximation accounts for Coulomb effects beyond the mean-field Hartree–Fock level, by taking into account all correlation terms up to quadratic order in the screened Coulomb interaction. This method allows one to describe interaction-induced dephasing due to carrier-carrier scattering and higher-order polarization interaction as well as the corresponding dynamic energy renormalizations beyond the Hartree–Fock level. When combined with a Markov approximation, i.e., neglecting memory effects, the correlation contributions reduce in the incoherent limit ($p_{\mathbf{k}}^{eh} \equiv 0$) to Boltzmann scattering integrals as well as generalized Coulomb enhancement and band-gap renormalization terms [21].

In the coherent regime, the second-order Born approximation can be used to study the dynamics of the coherent excitonic polarization and the coherently driven carrier population under the influence of carrier-carrier scattering and nonlinear polarization interaction. Unlike the dynamics-controlled truncation scheme, which is discussed below, the derived terms are not restricted to certain powers in the exciting field. However, the truncation in terms of powers of the screened Coulomb interaction excludes the description of bound biexciton states in the second-order Born approximation.

The derivation of all correlation contributions that arise in second-order Born–Markov approximation is straightforward but lengthy and can be found in [7, 11, 13, 14, 34]. Therefore, we restrict the present discussion to the most important steps using one particular correlation function as an example. Furthermore, for clarity we restrict the following analysis to the two-band case; for a discussion of multiband situations, see, e.g., [4]. Note, that for the 2×2 band situation with diagonal selection rules introduced above, the system reduces to a two-band situation when the incident fields have all the same circular polarization such that only one spin subsystem of the spin-degenerate electron and heavy-hole bands is optically excited.

Let us start by considering $\left\langle a_{e,\mathbf{k}}^{\dagger} a_{e,\mathbf{k}'-\mathbf{q}}^{\dagger} a_{e,\mathbf{k}-\mathbf{q}} a_{e,\mathbf{k}'} \right\rangle$, which is one term that appears when the electron occupation $f_{\mathbf{k}}^{ee}$ is commuted with the Coulomb interaction H_{Coulomb} [1]. It is now convenient to define a reduced four-operator correlation function by subtracting the uncorrelated Hartree–Fock part via

$$\Delta \left\langle a^{\dagger}_{e,\mathbf{k}} a^{\dagger}_{e,\mathbf{k'}-\mathbf{q}} a_{e,\mathbf{k}-\mathbf{q}} a_{e,\mathbf{k'}} \right\rangle = \left\langle a^{\dagger}_{e,\mathbf{k}} a^{\dagger}_{e,\mathbf{k'}-\mathbf{q}} a_{e,\mathbf{k}-\mathbf{q}} a_{e,\mathbf{k'}} \right\rangle$$
$$- \left\langle a^{\dagger}_{e,\mathbf{k}} a_{e,\mathbf{k}} \right\rangle \left\langle a^{\dagger}_{e,\mathbf{k}-\mathbf{q}} a_{e,\mathbf{k}-\mathbf{q}} \right\rangle \delta_{\mathbf{k},\mathbf{k'}} . \quad (18)$$

The dynamics of this correlated part of the four-operator correlation function is determined by the equation of motion:

$$\frac{\partial}{\partial t} \Delta \left\langle a^{\dagger}_{e,\mathbf{k}} a^{\dagger}_{e,\mathbf{k'}-\mathbf{q}} a_{e,\mathbf{k}-\mathbf{q}} a_{e,\mathbf{k'}} \right\rangle = \frac{\partial}{\partial t} \left\langle a^{\dagger}_{e,\mathbf{k}} a^{\dagger}_{e,\mathbf{k'}-\mathbf{q}} a_{e,\mathbf{k}-\mathbf{q}} a_{e,\mathbf{k'}} \right\rangle$$
$$- \left(\frac{\partial}{\partial t} \left\langle a^{\dagger}_{e,\mathbf{k}} a_{e,\mathbf{k}} \right\rangle \right) \left\langle a^{\dagger}_{e,\mathbf{k}-\mathbf{q}} a_{e,\mathbf{k}-\mathbf{q}} \right\rangle \delta_{\mathbf{k},\mathbf{k'}}$$
$$- \left\langle a^{\dagger}_{e,\mathbf{k}} a_{e,\mathbf{k}} \right\rangle \left(\frac{\partial}{\partial t} \left\langle a^{\dagger}_{e,\mathbf{k}-\mathbf{q}} a_{e,\mathbf{k}-\mathbf{q}} \right\rangle \right) \delta_{\mathbf{k},\mathbf{k'}} . \quad (19)$$

The time derivates on the right hand side of (19) can be obtained by evaluating the Heisenberg equation, (4). Performing the commutations, one finds that the resulting expression can be written as

$$\frac{\partial}{\partial t} \Delta \left\langle a^{\dagger}_{e,\mathbf{k}} a^{\dagger}_{e,\mathbf{k'}-\mathbf{q}} a_{e,\mathbf{k}-\mathbf{q}} a_{e,\mathbf{k'}} \right\rangle = \frac{\partial}{\partial t} \Delta \left\langle a^{\dagger}_{e,\mathbf{k}} a^{\dagger}_{e,\mathbf{k'}-\mathbf{q}} a_{e,\mathbf{k}-\mathbf{q}} a_{e,\mathbf{k'}} \right\rangle \Big|_{hom}$$
$$+ \frac{\partial}{\partial t} \Delta \left\langle a^{\dagger}_{e,\mathbf{k}} a^{\dagger}_{e,\mathbf{k'}-\mathbf{q}} a_{e,\mathbf{k}-\mathbf{q}} a_{e,\mathbf{k'}} \right\rangle \Big|_{inhom} , \quad (20)$$

with

$$i\hbar \frac{\partial}{\partial t} \Delta \left\langle a^{\dagger}_{e,\mathbf{k}} a^{\dagger}_{e,\mathbf{k'}-\mathbf{q}} a_{e,\mathbf{k}-\mathbf{q}} a_{e,\mathbf{k'}} \right\rangle \Big|_{hom} = - \left(-\varepsilon^{e}_{\mathbf{k}} - \varepsilon^{e}_{\mathbf{k'}-\mathbf{q}} + \varepsilon^{e}_{\mathbf{k}-\mathbf{q}} + \varepsilon^{e}_{\mathbf{k'}} \right)$$
$$\times \Delta \left\langle a^{\dagger}_{e,\mathbf{k}} a^{\dagger}_{e,\mathbf{k'}-\mathbf{q}} a_{e,\mathbf{k}-\mathbf{q}} a_{e,\mathbf{k'}} \right\rangle . \quad (21)$$

The homogeneous part of (20) simply contains the differences of kinetic energies. Its inhomogeneous part is given by the commutator with the Coulomb interaction, which is given by lengthy expressions which include coupling to six-operator terms.

In the second-order Born–Markov treatment one solves (20) adiabatically, by approximating

$$\Delta \left\langle a^{\dagger}_{e,\mathbf{k}} a^{\dagger}_{e,\mathbf{k'}-\mathbf{q}} a_{e,\mathbf{k}-\mathbf{q}} a_{e,\mathbf{k'}} \right\rangle \approx \frac{\frac{\partial}{\partial t} \Delta \left\langle a^{\dagger}_{e,\mathbf{k}} a^{\dagger}_{e,\mathbf{k'}-\mathbf{q}} a_{e,\mathbf{k}-\mathbf{q}} a_{e,\mathbf{k'}} \right\rangle \Big|_{inhom}}{\frac{i}{\hbar} \left(-\varepsilon^{e}_{\mathbf{k}} - \varepsilon^{e}_{\mathbf{k'}-\mathbf{q}} + \varepsilon^{e}_{\mathbf{k}-\mathbf{q}} + \varepsilon^{e}_{\mathbf{k'}} \right) - \gamma} ,$$
$$(22)$$

where γ is a decay constant. In the expressions that are given below, we use the limit $\gamma \to 0$. Since the Markov approximation has been used, memory

effects are ignored at this level. The second-order Born–Markov approximation corresponds to factorizing all four- and six-operator terms that appear
in the numerator of the right hand side of (22) into products of expectation
values of two operators, i.e., interband polarizations and carrier occupations.
Thus, the analysis is restricted to two-particle collisions, i.e., one keeps all
terms up to quadratic order in the screened Coulomb potential.

For a two-band situation, one obtains the equation of motion for the
microscopic polarization $p_\mathbf{k}^{eh}(t)$ as [13, 14]

$$\left[i\hbar\frac{\partial}{\partial t} - \varepsilon_\mathbf{k}^e(t) - \varepsilon_\mathbf{k}^h(t)\right] p_\mathbf{k}^{eh}(t) = \left[1 - f_\mathbf{k}^e - f_\mathbf{k}^h\right]\Omega_\mathbf{k}(t)$$

$$- i\,\Gamma_\mathbf{k}\, p_\mathbf{k}^{eh}(t) + i\sum_{\mathbf{k}'}\Gamma_{\mathbf{k},\mathbf{k}'}\, p_{\mathbf{k}+\mathbf{k}'}^{eh}(t)\,, \qquad (23)$$

where $\Omega_\mathbf{k}(t)$ and $\varepsilon_\mathbf{k}^a(t)$ with $(a = e, h)$ are the renormalized field and single-
particle energies which are defined in (13) and (14). Here, we use the short
notation for the electron and hole populations, i.e., $f_\mathbf{k}^e \equiv f_\mathbf{k}^{ee}$ and $f_\mathbf{k}^h \equiv f_\mathbf{k}^{hh}$.

Equation (23) contains correlation contributions which are diagonal in
the carrier momentum

$$\Gamma_\mathbf{k} = \sum_{\mathbf{k}',\mathbf{k}''}\sum_{a,b=e,h} g\left(\varepsilon_\mathbf{k}^a + \varepsilon_{\mathbf{k}'+\mathbf{k}''}^b - \varepsilon_{\mathbf{k}''}^b - \varepsilon_{\mathbf{k}'+\mathbf{k}}^a\right)$$

$$\times \left[W_{\mathbf{k}'}^2 - \delta_{ab}\,W_{\mathbf{k}'}W_{\mathbf{k}-\mathbf{k}''}\right]$$

$$\times \left[(1 - f_{\mathbf{k}'+\mathbf{k}''}^b)f_{\mathbf{k}''}^b f_{\mathbf{k}'+\mathbf{k}}^a + f_{\mathbf{k}'+\mathbf{k}''}^b(1 - f_{\mathbf{k}''}^b)(1 - f_{\mathbf{k}'+\mathbf{k}}^a)\right.$$

$$\left.-p_{\mathbf{k}'+\mathbf{k}''}^{eh}(p_{\mathbf{k}''}^{eh})^*\right]\,, \qquad (24)$$

and terms which couple different states $\mathbf{k}$ and $\mathbf{k}'$

$$\Gamma_{\mathbf{k},\mathbf{k}'} = \sum_{\mathbf{k}''}\sum_{a,b=e,h} g\left(-\varepsilon_\mathbf{k}^a - \varepsilon_{\mathbf{k}'+\mathbf{k}''}^b + \varepsilon_{\mathbf{k}''}^b + \varepsilon_{\mathbf{k}'+\mathbf{k}}^a\right)$$

$$\times \left[W_{\mathbf{k}'}^2 - \delta_{ab}\,W_{\mathbf{k}'}W_{\mathbf{k}-\mathbf{k}''}\right]$$

$$\times \left[(1 - f_\mathbf{k}^a)(1 - f_{\mathbf{k}'+\mathbf{k}''}^b)f_{\mathbf{k}''}^b + f_\mathbf{k}^a f_{\mathbf{k}'+\mathbf{k}''}^b(1 - f_{\mathbf{k}''}^b)\right.$$

$$\left.-(p_{\mathbf{k}'+\mathbf{k}''}^{eh})^* p_{\mathbf{k}''}^{eh}\right]\,, \qquad (25)$$

where $W_\mathbf{k}$ is the screened Coulomb interaction, which is evaluated in quasi-
static random-phase approximation (Lindhard formula), and $g(\varepsilon) = \pi\delta(\varepsilon) + \mathcal{P}\frac{i}{\varepsilon}$, where $\mathcal{P}$ denotes that the principal value of the corresponding integral
has to be taken.

The occupations for electrons and holes $f_\mathbf{k}^e(t)$ and $f_\mathbf{k}^h(t)$ obey the following
equation of motion

$$i\hbar\frac{\partial}{\partial t}f_\mathbf{k}^a(t) = -2\mathrm{Im}\left[\Omega_\mathbf{k}(t)(p_\mathbf{k}^{eh}(t))^*\right] \qquad (26)$$

$$+ i\left\{\Sigma_\mathbf{k}^{in,a}(t)\left[1 - f_\mathbf{k}^a(t)\right] - \Sigma_\mathbf{k}^{out,a}(t)f_\mathbf{k}^a(t) + \Sigma_\mathbf{k}^{pol,a}(t)\right\}\,.$$

The correlation contributions in (26) describe the changes of the carrier occupation probabilities $f_{\mathbf{k}}^a$ due to scattering of carriers into the state with momentum $\mathbf{k}$

$$
\begin{aligned}
\Sigma_{\mathbf{k}}^{in,a} = 2\pi \sum_{\mathbf{k}',\mathbf{k}''} \sum_{b=e,h} & \left[W_{\mathbf{k}'}^2 - \delta_{ab} \, W_{\mathbf{k}'} W_{\mathbf{k}-\mathbf{k}''} \right] \\
& \times \delta \left(\varepsilon_{\mathbf{k}}^a + \varepsilon_{\mathbf{k}'+\mathbf{k}''}^b - \varepsilon_{\mathbf{k}''}^b - \varepsilon_{\mathbf{k}'+\mathbf{k}}^a \right) \\
& \times (1 - f_{\mathbf{k}'+\mathbf{k}''}^b) f_{\mathbf{k}''}^b f_{\mathbf{k}'+\mathbf{k}}^a \,,
\end{aligned}
\tag{27}
$$

due to scattering of carriers out of the $\mathbf{k}$-state,

$$
\begin{aligned}
\Sigma_{\mathbf{k}}^{out,a} = 2\pi \sum_{\mathbf{k}',\mathbf{k}''} \sum_{b=e,h} & \left[W_{\mathbf{k}'}^2 - \delta_{ab} \, W_{\mathbf{k}'} W_{\mathbf{k}-\mathbf{k}''} \right] \\
& \times \delta \left(\varepsilon_{\mathbf{k}}^a + \varepsilon_{\mathbf{k}'+\mathbf{k}''}^b - \varepsilon_{\mathbf{k}''}^b - \varepsilon_{\mathbf{k}'+\mathbf{k}}^a \right) \\
& \times f_{\mathbf{k}'+\mathbf{k}''}^b (1 - f_{\mathbf{k}''}^b)(1 - f_{\mathbf{k}'+\mathbf{k}}^a) \,,
\end{aligned}
\tag{28}
$$

as well as due to nonlinear interactions with polarizations

$$
\begin{aligned}
\Sigma_{\mathbf{k}}^{pol,a} = \sum_{\mathbf{k}',\mathbf{k}''} \sum_{b=e,h} & \left[W_{\mathbf{k}'}^2 - \delta_{ab} \, W_{\mathbf{k}'} W_{\mathbf{k}-\mathbf{k}''} \right] \\
& \times \Big\{ g \left(\varepsilon_{\mathbf{k}}^a + \varepsilon_{\mathbf{k}'+\mathbf{k}''}^b - \varepsilon_{\mathbf{k}''}^b - \varepsilon_{\mathbf{k}'+\mathbf{k}}^a \right) \\
& \quad\quad \times \left[(f_{\mathbf{k}}^a - f_{\mathbf{k}'+\mathbf{k}}^a) \, p_{\mathbf{k}'+\mathbf{k}''}^{eh} (p_{\mathbf{k}''}^{eh})^* + c.c. \right] \\
& + \; g \left(\varepsilon_{\mathbf{k}}^{\bar{a}} + \varepsilon_{\mathbf{k}'+\mathbf{k}''}^b - \varepsilon_{\mathbf{k}''}^b - \varepsilon_{\mathbf{k}'+\mathbf{k}}^{\bar{a}} \right) \\
& \quad\quad \times \left[(f_{\mathbf{k}''}^b - f_{\mathbf{k}'+\mathbf{k}''}^b) \, p_{\mathbf{k}}^{eh} (p_{\mathbf{k}'+\mathbf{k}}^{eh})^* + c.c. \right] \Big\} \,.
\end{aligned}
\tag{29}
$$

More details can be found in, e.g., [14, 21].

4.2 Density-Dependent Exciton Saturation and Broadening

In the equation of motion for the microscopic polarization in second-order Born approximation, one recognizes the structure of a Boltzmann scattering integral. However, the polarization itself, as well as the rates Γ are complex quantities. The term with $\Gamma_{\mathbf{k}}$ on the right hand side of (23) can be considered as a microscopic expression for a diagonal dephasing rate, whereas the second term with $\Gamma_{\mathbf{k},\mathbf{k}'}$ represents non-diagonal dephasing. Diagonal and non-diagonal refer here to the matrix structure of the equation when the different $p_{\mathbf{k}}^{eh}$ are considered as a vector in $\mathbf{k}$-space and the right hand side of (23) as the product of the dephasing matrix times that vector. Solving this equation numerically, it turns out, that strong cancellation effects between the diagonal and non-diagonal contributions occur. As a consequence, one obtains incorrect estimates of the dephasing time if one ignores the off-diagonal parts.

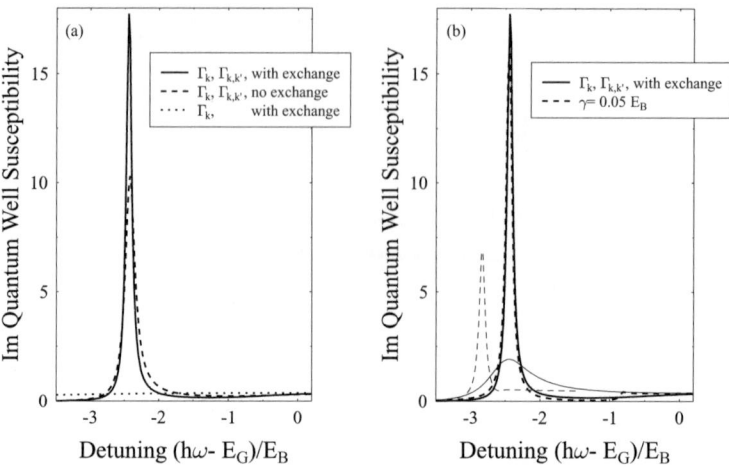

Fig. 2. (a) Computed excitonic absorption spectra (imaginary part of the optical susceptibility) using the parameters for an 8 nm GaAs-type quantum well and plasma density $10^{10}\,\mathrm{cm}^{-2}$ at $T = 77\,\mathrm{K}$. The *solid line* has been obtained with the full dephasing terms, the *dashed line* without exchange contribution to the Coulomb matrix element, the dotted line without the off-diagonal dephasing term in (23). (**b**) Comparison of the full microscopic dephasing calculation (*solid lines*) with computations assuming constant damping (*dashed lines*) for carrier density $10^{10}\,\mathrm{cm}^{-2}$ (*thick lines*) and $10^{11}\,\mathrm{cm}^{-2}$ (*thin lines*). Taken from [34]

The strong compensation among the diagonal and non-diagonal correlation terms is visualized in Fig. 2 by showing numerical results for the excitonic absorption in the presence of an incoherent electron-hole plasma. Figure 2a demonstrates that the full calculation yields a well-defined excitonic resonance. If, however, the off-diagonal terms are neglected, one finds an almost complete saturation already at the rather low plasma density of $10^{10}\,\mathrm{cm}^{-2}$ at $T = 77\,\mathrm{K}$. In Fig. 2b, the full result is compared to a calculation in which a constant dephasing time has been used. The constant dephasing approximation leads to artificial, density-dependent shifts of the saturating exciton resonance. Altogether these results demonstrate clearly the danger of describing the dephasing of the polarization at the simple level of T_2 times. At best, such times can be regarded as effective quantities that have been introduced empirically and which are valid only under some restrictive conditions. For a microscopic description of Coulomb-induced dephasing, it is necessary to evaluate the full polarization decay dynamics according to (23).

The dependence of the effective dephasing on the intensity of an optical excitation is demonstrated in Fig. 3. The dynamical evolution of the optical polarization after resonant excitation with a short fs optical pulse is shown in Fig. 3a. Besides the more rapid decay for higher excitation intensities, the appearing oscillations are quantum beats between the excitonic resonances

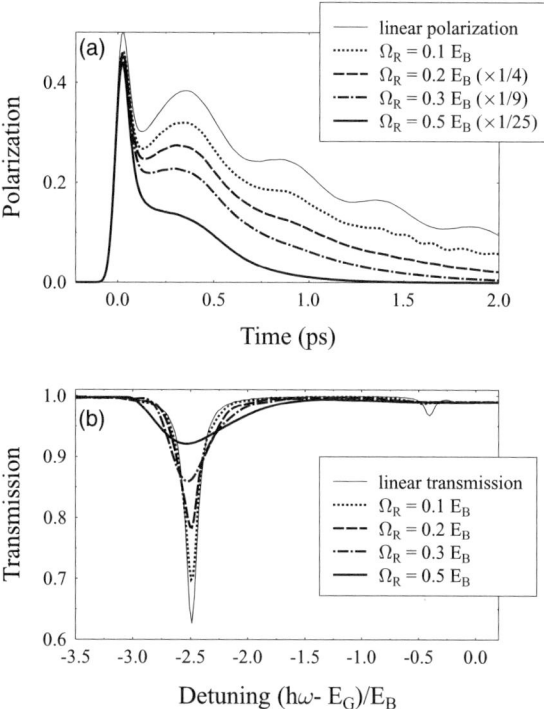

Fig. 3. (a) Time-dependent polarization and (b) transmission computed for an 8 nm GaAs single quantum-well structure. The curves are obtained assuming a 100 fs external laser pulse with increasing Rabi energy Ω_R (given in units of the exciton binding energy E_B). Taken from [14]

of the system. Figure 3b depicts the concomitant gradual saturation and broadening of the exciton resonance in the absorption spectra. This behavior is caused by *excitation-induced dephasing*, which is the result of the dependence of $\Gamma_\mathbf{k}$ and $\Gamma_{\mathbf{k},\mathbf{k}'}$ on the time-dependent population and polarization. The fact that the $\mathbf{k}$ sum of the scattering rates Γ vanishes [14], indicates that Coulomb dephasing of the optical polarization is a consequence of destructive interference between microscopic polarization contributions [15].

Experimentally, excitation-induced dephasing was first identified in four-wave-mixing studies which were performed with pre-pulses [35]. The polarization-dependent decay of the four-wave-mixing decay has been analyzed [36] and excitonic saturation and broadening has been observed in pump-probe configuration using resonant short pulse excitation, see e.g., [21, 37]. The main qualitative features observed experimentally are in nice agreement with the microscopic theory.

From a slightly different perspective, the microscopic analysis of dephasing is nothing but a detailed line shape theory. The second-order Born–Markov approximation has successfully been applied to predict optical spectra in systems with elevated electron-hole-plasma densities, where part of the absorption becomes negative, i.e., optical gain is realized. Such highly excited gain media are used in semiconductor lasers. The calculation of the proper gain line shape of such structures has been a long standing problem since the use of a constant dephasing approximation leads to the presence of unphysical absorption appearing energetically below the gain region [38]. As it turns out, the numerical solutions of the semiconductor Bloch equations, where dephasing is treated according to (23), yield a solution of the laser line shape problem. Such evaluations of the optical spectra often lead to a very good agreement with experimental results. More background and details of the microscopic semiconductor laser and gain theory, as well as many numerical results, can be found in [38] and references cited therein.

4.3 Dynamics-Controlled Truncation Scheme: Coherent $\chi^{(3)}$-Limit

Instead of using many-body methods, which are nonperturbative in the fields, to close the dynamic equations for the reduced single-particle density matrix, (3), one can alternatively follow an approach where the nonlinear optical response is classified according to an expansion in powers of the optical field. Stahl and coworkers [10] were the first who recognized that this traditional nonlinear optics expansion [22] introduces a systematic truncation scheme for the many-body Coulomb correlations in the case of purely coherent optical excitation configurations. In the following, we outline the basic steps that are involved in this dynamics-controlled truncation procedure. More details can be found in [9, 10, 12, 18, 31].

Studying the structure of the coupled equations for the correlation functions, one finds that the four-operator terms which appear in the equation of motion for the microscopic polarization can be expressed according to [9, 18]

$$\left\langle a^\dagger_{e,\mathbf{k}} a^\dagger_{e',\mathbf{k}'} b^\dagger_{h'',\mathbf{k}''} a_{e''',\mathbf{k}'''} \right\rangle = \sum_{\hat{\mathbf{k}},\hat{h}} \left\langle a^\dagger_{e,\mathbf{k}} a^\dagger_{e',\mathbf{k}'} b^\dagger_{h'',\mathbf{k}''} b^\dagger_{\hat{h},\hat{\mathbf{k}}} \right\rangle \left\langle b_{\hat{h},\hat{\mathbf{k}}} a_{e''',\mathbf{k}'''} \right\rangle$$

$$+ O(E^5) \,, \tag{30}$$

where $\left\langle a^\dagger_{e,\mathbf{k}} a^\dagger_{e',\mathbf{k}'} b^\dagger_{h'',\mathbf{k}''} b^\dagger_{\hat{h},\hat{\mathbf{k}}} \right\rangle$ can be considered as an unfactorized product of two polarization operators that is at least of second order in the field, $\propto O(E^2)$, since two electron-hole pairs are created, whereas $\left\langle b_{\hat{h},\hat{\mathbf{k}}} a_{e''',\mathbf{k}'''} \right\rangle$ is already present in equations that are linear in the field, $\propto O(E)$, since a single electron-hole pair is destroyed. Hence, the right hand side of (30) is at least of third order in the field. As is outlined below, by evaluating the equations of motions for the terms appearing in (30) to the lowest order in the field, it is quite easy to verify the correctness of this decoupling scheme. As shown afterwards, however, one needs more general considerations to see where this expression originates from and how to analyze other, more complicated expectations values.

To verify (30), one first notes that it is valid when the semiconductor is in its ground state, since in this case both sides vanish. Then one can take the time derivative of the lowest (third) order contributions of (30) which leads to

$$
\left\langle \frac{\partial}{\partial t} (a^\dagger_{e,\mathbf{k}} a^\dagger_{e',\mathbf{k}'} b^\dagger_{h'',\mathbf{k}''} a_{e''',\mathbf{k}'''}) \right\rangle
$$

$$
= \sum_{\hat{\mathbf{k}},\hat{h}} \left\langle \frac{\partial}{\partial t} (a^\dagger_{e,\mathbf{k}} a^\dagger_{e',\mathbf{k}'} b^\dagger_{h'',\mathbf{k}''} b^\dagger_{\hat{h},\hat{\mathbf{k}}}) \right\rangle \left\langle b_{\hat{h},\hat{\mathbf{k}}} a_{e''',\mathbf{k}'''} \right\rangle
$$

$$
+ \sum_{\hat{\mathbf{k}},\hat{h}} \left\langle a^\dagger_{e,\mathbf{k}} a^\dagger_{e',\mathbf{k}'} b^\dagger_{h'',\mathbf{k}''} b^\dagger_{\hat{h},\hat{\mathbf{k}}} \right\rangle \left\langle \frac{\partial}{\partial t} (b_{\hat{h},\hat{\mathbf{k}}} a_{e''',\mathbf{k}'''}) \right\rangle . \qquad (31)
$$

Computing the time derivatives by evaluating the commutators with the Hamiltonian according to the Heisenberg equation, inserting the obtained equations of motion up to the required order in the field, and performing the summations, (31) and thus (30) are readily verified.

Another, even simpler example for a decoupling scheme which is valid in a fully coherent situation is the expression for the occupation probabilities in terms of the microscopic polarizations up to second order in the field

$$
f^{ee'}_{\mathbf{k}} = \sum_{h'} (p^{eh'}_{\mathbf{k}})^* p^{e'h'}_{\mathbf{k}} + O(E^4) ,
$$

$$
f^{hh'}_{\mathbf{k}} = \sum_{e'} (p^{e'h}_{\mathbf{k}})^* p^{e'h'}_{\mathbf{k}} + O(E^4) . \qquad (32)
$$

Analogously to what has been sketched above, also this conservation law, which is well known from a two-level system, can be verified easily by computing its time derivative and inserting the expressions up to the required order in the field.

Besides just confirming the decoupling schemes given above, one can also address the coherent dynamics of a many-body system on the level of a general analysis. As an example, we discuss in the following the two conservation laws given above as special cases and show how it is possible to generalize such expressions. Let us start by defining a normally ordered operator product of creation and annihilation as [9]

$$
\{N, M\} \equiv c^\dagger(\alpha_N) c^\dagger(\alpha_{N-1}) \ldots c^\dagger(\alpha_1) c(\beta_1) \ldots c(\beta_{M-1}) c(\beta_M) , \qquad (33)
$$

where depending on α_i and β_j the operators $c^\dagger$ and c are electron or hole operators for certain $\mathbf{k}_i$ and $\mathbf{k}_j$, respectively. The quantities $\{N, M\}$ contain the full information about the temporal evolution of the photoexcited many-body system. E.g., the microscopic polarization $p^{eh}_{\mathbf{k}}$ corresponds to $\{0, 2\}$, its complex conjugate $(p_{\mathbf{k}}^{eh})^*$ to $\{2,0\}$, and the carrier occupation probabilities $f_{\mathbf{k}}^{aa}$ to $\{1,1\}$.

We now consider the time derivative of the normally ordered operator products, which is given by

$$\frac{\partial}{\partial t}\{N+1,M+1\} = \left(\frac{\partial c^{\dagger}(\alpha_{N+1})}{\partial t}\right)\{N,M\}\,c(\beta_{M+1})$$
$$+\,c^{\dagger}(\alpha_{N+1})\left(\frac{\partial\{N,M\}}{\partial t}\right)\,c(\beta_{M+1})$$
$$+\,c^{\dagger}(\alpha_{N+1})\{N,M\}\left(\frac{\partial c(\beta_{M+1})}{\partial t}\right). \tag{34}$$

To analyze to which correlations functions $\{N,M\}$ is coupled, one can evaluate the equations of motion for the individual operators, i.e., $\frac{\partial c^{\dagger}(\alpha_i)}{\partial t}$ and $\frac{\partial c(\beta_j)}{\partial t}$. By commuting these operators with the Hamiltonian one finds that $\{1,0\}$ ($\{0,1\}$) is coupled to itself by the single-particle (band structure) part, coupled to $\{0,1\}$ ($\{1,0\}$) by the light-matter interaction, and coupled to $\{2,1\}$ ($\{1,2\}$) by the many-body Coulomb interaction [9].

By inserting these results into (34) and restoring normal ordering, one finds that the light-matter interaction couples $\{N,M\}$ to $\{N-1,M+1\}$, $\{N+1,M-1\}$, $\{N-2,M\}$, and $\{N,M-2\}$, since it is given by a single pair of one creation and one destruction operator. The many-body Coulomb interaction, however, couples $\{N,M\}$ to itself and to $\{N+1,M+1\}$ [9]. Thus, one verifies once again that it is the four-operator part of the Hamiltonian, which generates the coupling to products that contain more operators, i.e., the many-body hierarchy. Clearly, for both parts of the Hamiltonian, operators with $N-M$ being odd are only coupled to other operators where $N-M$ is odd and operators with $N-M$ being even are only coupled to other operators where $N-M$ is even. If we start from the ground state as the initial condition, i.e., $\langle\{0,0\}\rangle = 1$ and all other expectation values are vanishing, all operators $\{N,M\}$ with $N-M$ being odd vanish at all times, since their equations of motion contain no sources. The above considerations are visualized in Fig. 4. This figure clearly shows that in any finite order in the external field, only a finite number of expectation values contribute to the optical response [9, 10]. To find the lowest order in the light field at which any term $\{N,M\}$ is nonvanishing, one only needs to evaluate the minimum number of dotted lines in Fig. 4, which are needed to connect it to $\{0,0\}$. As one can see from Fig. 4 and as can be proven rigorously [9], the minimum order in the field in which $\langle\{N,M\}\rangle$ is finite is $(N+M)/2$ if N and M are both even and $(N+M)/2+1$ if N and M are both odd, i.e.,

$$\langle c^{\dagger}(\alpha_{2N})\ldots c^{\dagger}(\alpha_1)\,c(\beta_1)\ldots c(\beta_{2M})\rangle = O(E^{N+M}), \tag{35}$$

and

$$\langle c^{\dagger}(\alpha_{2N+1})\ldots c^{\dagger}(\alpha_1)\,c(\beta_1)\ldots c(\beta_{2M+1})\rangle = O(E^{N+M+2}). \tag{36}$$

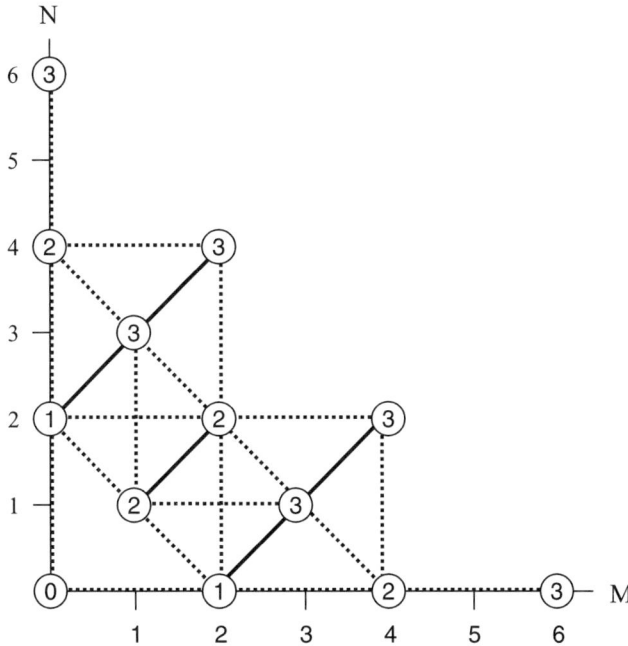

Fig. 4. Schematical drawing of relevant dynamic variables $\langle\{N, M\}\rangle$. The numbers in the symbols correspond to the minimum order in the external field in which $\langle\{N, M\}\rangle$ is finite. The *dotted lines* denote the couplings due to the light-matter interaction, whereas the *solid lines* indicate the couplings via the Coulomb interaction. After [9]

For the case where one initially starts from the ground state of the semiconductor and where the dynamics is fully coherent, one can factorize expectation values which contain a mix of creation and destruction operators to lowest order into products of expectation values which contain either only creation or only destruction operators. For the relevant combinations of operators in our case, we can write

$$\langle c^\dagger(\alpha_{2N}) \dots c^\dagger(\alpha_1)\, c(\beta_1) \dots c(\beta_{2M})\rangle$$
$$= \langle c^\dagger(\alpha_{2N}) \dots c^\dagger(\alpha_1)\rangle\, \langle c(\beta_1) \dots c(\beta_{2M})\rangle + O(E^{N+M+2}), \quad (37)$$

and

$$\langle c^\dagger(\alpha_{2N+1}) \dots c^\dagger(\alpha_1)\, c(\beta_1) \dots c(\beta_{2M+1})\rangle$$
$$= \sum_{\delta_1} \langle c^\dagger(\alpha_{2N+1}) \dots c^\dagger(\alpha_1) c^\dagger(\delta_1)\rangle\, \langle c(\delta_1) c(\beta_1) \dots c(\beta_{2M+1})\rangle$$
$$+ O(E^{N+M+4}). \quad (38)$$

To prove (37) and (38), it is important to note that in the Heisenberg picture the operators are time dependent and all expectation values are taken with respect to the initial state, which is the ground state ($|0\rangle$) of the semiconductor, i.e., [9]

$$\langle c^\dagger(\alpha_N)\, c^\dagger(\alpha_{N-1}) \dots c^\dagger(\alpha_1)\, c(\beta_1) \dots c(\beta_M)\rangle$$
$$= \langle 0 \,|\, c_H^\dagger(\alpha_N, t) \dots c_H^\dagger(\alpha_1, t)\, c_H(\beta_1, t) \dots c_H(\beta_M, t)\,|\, 0\rangle. \qquad (39)$$

Now, one can define an interaction picture by decomposing the Hamiltonian $H = H_{\text{single-particle}} + H_{\text{Coulomb}} + H_{\text{light-matter}}$, see (5), into two parts, via $H = H_0 + H_{\text{light-matter}}$, where we introduced $H_0 = H_{\text{single-particle}} + H_{\text{Coulomb}}$. In the interaction picture an operator $\mathcal{A}$ becomes time dependent as described by

$$\mathcal{A}_I(t) = e^{iH_0 t}\, \mathcal{A}\, e^{-iH_0 t}. \qquad (40)$$

By inserting the interaction representation of the identity operator in between the creation and destruction operators of a ground state expectation value of Heisenberg operators, i.e., in between $c_H^\dagger(\alpha_1, t)$ and $c_H(\beta_1, t)$ of (39), and iterating the exponentials $e^{\pm iH_0 t}$ of (40) [9], one obtains

$$\langle 0 \,|\, c_H^\dagger(\alpha_N, t) \dots c_H^\dagger(\alpha_1, t)\, c_H(\beta_1, t) \dots c_H(\beta_M, t)\,|\, 0\rangle$$
$$= \langle 0 \,|\, c_H^\dagger(\alpha_N, t) \dots c_H^\dagger(\alpha_1, t)\,|\, 0\rangle \langle 0 \,|\, c_H(\beta_1, t) \dots c_H(\beta_M, t)\,|\, 0\rangle$$
$$+ \sum_{\delta_1} \langle 0 \,|\, c_H^\dagger(\alpha_N, t) \dots c_H^\dagger(\alpha_1, t)\, c_I^\dagger(\delta_1, t)\,|\, 0\rangle$$
$$\times \langle 0 \,|\, c_I(\delta_1, t)\, c_H(\beta_1, t) \dots c_H(\beta_M, t)\,|\, 0\rangle$$
$$+ \frac{1}{2} \sum_{\delta_1, \delta_2} \langle 0 \,|\, c_H^\dagger(\alpha_N, t) \dots c_H^\dagger(\alpha_1, t)\, c_I^\dagger(\delta_2, t)\, c_I^\dagger(\delta_1, t)\,|\, 0\rangle$$
$$\times \langle 0 \,|\, c_I(\delta_1, t)\, c_I(\delta_2, t)\, c_H(\beta_1, t) \dots c_H(\beta_M, t)\,|\, 0\rangle$$
$$+ \cdots. \qquad (41)$$

This general decoupling scheme can be used to verify the two special cases which are relevant for the coherent $\chi^{(3)}$-limit, i.e., (30) and (32). The expressions given above allow one, furthermore, to extend the dynamics-controlled truncation scheme to higher orders in the field. As an example, we discuss in Sect. 4.5 the equations of motion for the coherent $\chi^{(5)}$-limit schematically and introduce some results obtained on this level in Sect. 4.6.

In the following, we show how the dynamics-controlled truncation scheme can be applied to obtain the equations of motion which describe the optical semiconductor response in the coherent $\chi^{(3)}$-limit. In order to be able to distinguish between the uncorrelated Hartree–Fock part and the correlation contributions, it is convenient to introduce a pure four-particle correlation function which is defined by

$$\bar{B}^{eh'e'h}_{\mathbf{k},\mathbf{k}',\mathbf{k}'',\mathbf{k}'''} = \left\langle a^{\dagger}_{e,\mathbf{k}} b^{\dagger}_{h',\mathbf{k}'} a^{\dagger}_{e',\mathbf{k}''} b^{\dagger}_{h,\mathbf{k}'''} \right\rangle$$

$$- \left\langle a^{\dagger}_{e,\mathbf{k}} b^{\dagger}_{h',\mathbf{k}'} \right\rangle \left\langle a^{\dagger}_{e',\mathbf{k}''} b^{\dagger}_{e,\mathbf{k}'''} \right\rangle - \left\langle a^{\dagger}_{e,\mathbf{k}} b^{\dagger}_{h,\mathbf{k}'''} \right\rangle \left\langle a^{\dagger}_{e',\mathbf{k}''} b^{\dagger}_{h',\mathbf{k}'} \right\rangle$$

$$= B^{eh'e'h}_{\mathbf{k},\mathbf{k}',\mathbf{k}'',\mathbf{k}'''}$$

$$- \left\langle a^{\dagger}_{e,\mathbf{k}} b^{\dagger}_{h',\mathbf{k}'} \right\rangle \left\langle a^{\dagger}_{e',\mathbf{k}''} b^{\dagger}_{e,\mathbf{k}'''} \right\rangle - \left\langle a^{\dagger}_{e,\mathbf{k}} b^{\dagger}_{h,\mathbf{k}'''} \right\rangle \left\langle a^{\dagger}_{e',\mathbf{k}''} b^{\dagger}_{h',\mathbf{k}'} \right\rangle . \quad (42)$$

Using the above introduced expansion scheme up to third order in the field, the polarization equation can be written as [31]

$$\frac{\partial}{\partial t} p^{eh}_{\mathbf{k}} = \frac{\partial}{\partial t} p^{eh}_{\mathbf{k}}\big|_{hom} + \sum_{n=1}^{3} \frac{\partial}{\partial t} p^{eh}_{\mathbf{k}}\big|_{inhom,n} , \quad (43)$$

with

$$i\hbar \frac{\partial}{\partial t} p^{eh}_{\mathbf{k}}\big|_{hom} = - \left(\varepsilon^{e}_{\mathbf{k}} + \varepsilon^{h}_{\mathbf{k}} \right) p^{eh}_{\mathbf{k}} + \sum_{\mathbf{q}} V_{\mathbf{q}} p^{eh}_{\mathbf{k}-\mathbf{q}} , \quad (44)$$

$$i\hbar \frac{\partial}{\partial t} p^{eh}_{\mathbf{k}}\big|_{inhom,1} = \left(\boldsymbol{\mu}^{eh} - \sum_{e',h'} (p^{eh'}_{\mathbf{k}})^{*} p^{e'h'}_{\mathbf{k}} \boldsymbol{\mu}^{e'h} \right.$$

$$\left. - \sum_{h',e'} \boldsymbol{\mu}^{eh'} (p^{e'h'}_{\mathbf{k}})^{*} p^{e'h}_{\mathbf{k}} \right) \cdot \mathbf{E} , \quad (45)$$

$$i\hbar \frac{\partial}{\partial t} p^{eh}_{\mathbf{k}}\big|_{inhom,2} = - \sum_{\mathbf{q},e',h'} V_{\mathbf{q}} \left[p^{eh'}_{\mathbf{k}} \left(p^{e'h'}_{\mathbf{k}} \right)^{*} p^{e'h}_{\mathbf{k}-\mathbf{q}} - p^{eh'}_{\mathbf{k}+\mathbf{q}} \left(p^{e'h'}_{\mathbf{k}+\mathbf{q}} \right)^{*} p^{e'h}_{\mathbf{k}} \right.$$

$$\left. + p^{eh'}_{\mathbf{k}+\mathbf{q}} \left(p^{e'h'}_{\mathbf{k}} \right)^{*} p^{e'h}_{\mathbf{k}} - p^{eh'}_{\mathbf{k}} \left(p^{e'h'}_{\mathbf{k}-\mathbf{q}} \right)^{*} p^{e'h}_{\mathbf{k}-\mathbf{q}} \right] , \quad (46)$$

$$i\hbar \frac{\partial}{\partial t} p^{eh}_{\mathbf{k}}\big|_{inhom,3} = \sum_{\mathbf{q},\mathbf{k}',e',h'} V_{\mathbf{q}} \left(p^{e'h'}_{\mathbf{k}'} \right)^{*} \left[\bar{B}^{eh'e'h}_{\mathbf{k},\mathbf{k}',\mathbf{k}'-\mathbf{q},\mathbf{k}-\mathbf{q}} - \bar{B}^{eh'e'h}_{\mathbf{k}+\mathbf{q},\mathbf{k}',\mathbf{k}'-\mathbf{q},\mathbf{k}} \right.$$

$$\left. + \bar{B}^{eh'e'h}_{\mathbf{k}+\mathbf{q},\mathbf{k}'+\mathbf{q},\mathbf{k}',\mathbf{k}} - \bar{B}^{eh'e'h}_{\mathbf{k},\mathbf{k}'+\mathbf{q},\mathbf{k}',\mathbf{k}-\mathbf{q}} \right] . \quad (47)$$

The homogeneous part of (43) is given by the kinetic energies of electrons and holes plus their Coulomb attraction, see (44). This term is diagonal in the basis of exciton eigenstates. The contributions denoted with the subscript *inhom* in (43) are the different inhomogeneous driving terms, which appear as the sources for the microscopic polarization. The direct coupling of the carrier system to the optical field is represented by the terms proportional to $\boldsymbol{\mu} \cdot \mathbf{E}$, see (45). These terms include the linear optical coupling ($\boldsymbol{\mu} \cdot \mathbf{E}$) and contributions which are a consequence of phase-space filling ($\boldsymbol{\mu} \cdot \mathbf{E} \, p^{*}p$).

Due to the many-body Coulomb interaction, (43) contains further optical nonlinearities. The contributions that are of first-order in the Coulomb interaction (Vpp^*p) are given by (46). Those, together with the phase-space filling terms, correspond to the Hartree–Fock limit of (43). The correlation contributions to the polarization equation, see (47), consist of four terms. Schematically, these terms have the structure $Vp^*\bar{B}$, where $\bar{B}$ is the genuine four-particle correlation function defined in (42). This clearly shows, that due to the many-body Coulomb interaction, the two-particle electron-hole amplitude p is coupled to higher-order correlation functions $\bar{B}$, which again is just the beginning of the general many-body hierarchy mentioned earlier. Due to the additivity of the three types of nonlinear sources in (43) the total nonlinear optical polarization can be written as the sum of the three contributions discussed above [31, 39, 40].

The Hartree–Fock part of (43) can be obtained by an expansion of (12), (13), and (14) up to third-order in the optical field and by making use of the conservation law, (32), which is valid for a coherent system. This procedure allows one to identify the above described first-order Coulomb terms as the third-order parts of the renormalized field and transition energy.

Clearly, if the calculation of the coherent nonlinear optical response of a semiconductor starts initially from the ground state and is restricted to a finite order in the applied field, the many-body hierarchy is truncated automatically and one is left with a finite number of correlation functions [9, 10, 12, 18]. This truncation occurs since one considers here purely optical excitation, where only the optical field exists as a linear source which in first-order induces a linear polarization, in second order leads to carrier occupations, and so on.

In more general situations, where incoherent populations may be present, a classification of the nonlinear optical response in terms of powers of a field is no longer meaningful. To analyze such situations, one can use a cluster expansion [28]. This scheme generalizes both the second-order Born and the dynamics-controlled truncation approaches and includes exciton occupations as well. For the purpose of this review, we do not consider exciton population effects, which have been the focus of a number of recent studies [28, 41], but restrict the analysis to the level provided by the second-order Born and the dynamics-controlled truncation schemes.

Equation (43) has been obtained within the coherent $\chi^{(3)}$-limit. In this limit, the nonlinear optical response is fully determined by the dynamics of single and two electron-hole-pair excitations, p and $\bar{B}$, respectively [9, 10, 12]. The equation for $\bar{B}$ is given by [31]

$$\frac{\partial}{\partial t}\bar{B}^{eh'e'h}_{\mathbf{k},\mathbf{k}',\mathbf{k}'',\mathbf{k}'''} = \frac{\partial}{\partial t}\bar{B}^{eh'e'h}_{\mathbf{k},\mathbf{k}',\mathbf{k}'',\mathbf{k}'''}\big|_{hom} + \frac{\partial}{\partial t}\bar{B}^{eh'e'h}_{\mathbf{k},\mathbf{k}',\mathbf{k}'',\mathbf{k}'''}\big|_{inhom} , \qquad (48)$$

with

$$i\hbar\frac{\partial}{\partial t}\,\bar{B}^{eh'e'h}_{\mathbf{k},\mathbf{k}',\mathbf{k}'',\mathbf{k}'''}\Big|_{hom} = -\left(\varepsilon^e_{\mathbf{k}} + \varepsilon^{h'}_{\mathbf{k}'} + \varepsilon^{e'}_{\mathbf{k}''} + \varepsilon^h_{\mathbf{k}'''}\right)\bar{B}^{eh'e'h}_{\mathbf{k},\mathbf{k}',\mathbf{k}'',\mathbf{k}'''}$$

$$+ \sum_{\mathbf{q}'} V_{q'}\left[\bar{B}^{eh'e'h}_{\mathbf{k}+\mathbf{q}',\mathbf{k}'+\mathbf{q}',\mathbf{k}'',\mathbf{k}'''} - \bar{B}^{eh'e'h}_{\mathbf{k}+\mathbf{q}',\mathbf{k}',\mathbf{k}''-\mathbf{q}',\mathbf{k}'''}\right.$$

$$+\bar{B}^{eh'e'h}_{\mathbf{k}+\mathbf{q}',\mathbf{k}',\mathbf{k}'',\mathbf{k}'''+\mathbf{q}'} + \bar{B}^{eh'e'h}_{\mathbf{k},\mathbf{k}'+\mathbf{q}',\mathbf{k}''+\mathbf{q}',\mathbf{k}'''}$$

$$\left.-\bar{B}^{eh'e'h}_{\mathbf{k},\mathbf{k}'+\mathbf{q}',\mathbf{k}'',\mathbf{k}'''-\mathbf{q}'} + \bar{B}^{eh'e'h}_{\mathbf{k},\mathbf{k}',\mathbf{k}''+\mathbf{q}',\mathbf{k}'''+\mathbf{q}'}\right]\,,\quad (49)$$

$$i\hbar\frac{\partial}{\partial t}\,\bar{B}^{eh'e'h}_{\mathbf{k},\mathbf{k}',\mathbf{k}'',\mathbf{k}'''}\Big|_{inhom} = -V_{|\mathbf{k}-\mathbf{k}'''|}\left(p^{eh}_{\mathbf{k}'''} - p^{eh}_{\mathbf{k}}\right)\left(p^{e'h'}_{\mathbf{k}'} - p^{e'h'}_{\mathbf{k}''}\right)$$

$$+V_{|\mathbf{k}-\mathbf{k}'|}\left(p^{eh'}_{\mathbf{k}'} - p^{eh'}_{\mathbf{k}}\right)\left(p^{e'h}_{\mathbf{k}'''} - p^{e'h}_{\mathbf{k}''}\right)\,.\quad (50)$$

Equation (49) includes in its homogeneous part the kinetic energies as well as the attractive and repulsive interactions between two electrons and two holes, i.e., the *biexciton problem*. The sources for the dynamics of $\bar{B}$ consist of the Coulomb interaction potential times nonlinear polarization terms (Vpp), see (50).

The third-order limit of the second-order Born approximation (see previous section) can be obtained from (48) if one neglects the Coulomb interactions in (49). In this approximation, bound biexcitons are ignored and the many-body correlations are described up to second order in the Coulomb interaction.

So called Coulomb memory effects are automatically included in the analysis if one solves the coupled set of (43) and (48). The appearance of the memory can be seen most easily by integrating (48) with the schematical solution

$$\bar{B}(t) = \frac{i}{\hbar}V\int_{-\infty}^{t} dt'\,e^{i\bar{\varepsilon}(t-t')/\hbar}p(t')p(t')\,.\quad (51)$$

Inserting this expression for $\bar{B}(t)$ into (43) shows clearly that $p(t)$ depends on contributions $p(t')$ with $t' < t$, i.e., on terms with earlier time arguments. Such memory effects are neglected if one solves (48) adiabatically within the Markov approximation.

When comparing the dynamics-controlled truncation approach with the second-order Born approximation for the Coulomb correlation contributions, it should be noted that the second-order Born approximation is generally neither restricted to the coherent regime nor to a finite order in the interaction with the field, and, furthermore, also takes into account the screening of the Coulomb interaction. However, since the second-order Born approximation does not include exciton populations and bound biexcitons, it is particularly well suited to describe excitation conditions where electron-hole plasma effects dominate the nonlinear response. The third-order dynamics-controlled truncation scheme, on the other hand, describes the dynamics of bound and unbound biexciton resonances explicitly by following the evolution of the four-particle correlation function $\bar{B}$. It is therefore well suited to analyze the influence of such biexciton correlations on the nonlinear optical semiconductor response in coherent situations for weak excitation intensities.

Since the equations of motion in the coherent $\chi^{(3)}$-limit are quite lengthy, it is, for the clarity of discussions, sometimes useful to consider a reduced schematical set equations of motion. Such equations can be obtained from (43) and (48) by neglecting all indices and superscripts and the vector character, and have the following structure [39]

$$
\begin{aligned}
i\hbar \frac{\partial}{\partial t} p = &- \varepsilon_p\, p \\
&+ (\mu - b\, p^*\, p)\, E \\
&- V_{HF}\, p^*\, p\, p \\
&+ V_{corr}\, p^*\, \bar{B}
\end{aligned}
\tag{52}
$$

and

$$
\begin{aligned}
i\hbar \frac{\partial}{\partial t} \bar{B} = &- \varepsilon_B\, \bar{B} \\
&- \bar{V}_{corr}\, p\, p .
\end{aligned}
\tag{53}
$$

Here, ε_p and ε_B are the energies of a single- and a biexciton state, respectively, b denotes the Pauli blocking, V_{HF} the first-order (Hartree–Fock) Coulomb terms, and V_{corr} as well as $\bar{V}_{corr}$ Coulomb correlation contributions. If one uses instead of $\bar{B}$, the full four-point correlation function B given by $\bar{B}$ plus the Hartree–Fock contribution, see (42), one obtains [39]

$$
\begin{aligned}
i\hbar \frac{\partial}{\partial t} p = &- \varepsilon_p\, p \\
&+ (\mu - b\, p^*\, p)\, E \\
&+ V_{corr}\, p^*\, B
\end{aligned}
\tag{54}
$$

and

$$
\begin{aligned}
i\hbar \frac{\partial}{\partial t} B = &- \varepsilon_B\, B \\
&- \tilde{\mu}\, p\, E .
\end{aligned}
\tag{55}
$$

Here, $\tilde{\mu}$ is the optical matrix element for the exciton to biexciton transition.

4.4 Signatures of Coherent Four-Particle Correlations in $\chi^{(3)}$

On the basis of the coherent $\chi^{(3)}$-equations, one can analyze a variety of optical experiments such as low intensity pump probe and four-wave mixing, see, e.g., [2, 3, 4, 31, 39, 40, 42, 43, 44, 45, 46, 47, 48, 49] and references therein. To get a feeling for the contributions induced by four-particle correlations, we focus in the following on low intensity pump-probe configurations considering pumping either resonantly at or spectrally below the exciton. For simplicity, we assume here that the energetic splitting between the heavy-hole and light-hole excitons is much larger than the spectral width of the pump pulse and its possible detuning to the heavy-hole exciton which allows us to neglect the light-hole transitions. Results for which the light-hole transitions are relevant have been reported in, e.g., [3, 47, 48, 49, 50].

To keep the numerical requirements within reasonable limits, we have calculated the differential absorption for a one-dimensional tight-binding model system [39, 46]. Homogeneous broadening is treated phenomenologically by adding the decay rates $\gamma_p = 1/3$ ps and $\gamma_B = 1/1.5$ ps to the equations of motion (43) and (48), respectively. As shown in [46], the differential absorption spectra obtained from the one-dimensional model system are in qualitative agreement with two-dimensional calculations and experiments performed on high-quality quantum wells.

In the following analysis, we write the differential absorption spectrum as measured in a pump-probe experiment as the sum of three contributions [31, 39, 40]

$$\delta\alpha(\omega) = \delta\alpha_{pb}(\omega) + \delta\alpha_{CI,1st}(\omega) + \delta\alpha_{CI,corr}(\omega). \tag{56}$$

Here, pb denotes the optical nonlinearity induced by Pauli blocking. The terms denoted with CI are due to Coulomb-interaction-induced nonlinearities. $CI, 1st$ is the first-order (Hartree–Fock) term, and $CI, corr$ represents the higher-order correlation contributions.

The calculated differential absorption spectrum for resonant excitation at the exciton resonance with co-circularly polarized pump and probe pulses is shown in Fig. 5a. $\delta\alpha(\omega)$ is negative in the vicinity of the exciton resonance corresponding to bleaching of the exciton due to the pump-induced optical excitations. For energies greater than the exciton energy, we see in Fig. 5a positive contributions to $\delta\alpha(\omega)$, which are explained below. In addition to the total signal in the upper part of Fig. 5a, the three individual contributions to $\delta\alpha(\omega)$, see (56), arising from Pauli blocking, as well as first- and higher-order Coulomb contributions are displayed separately in the lower part. One can see that $\delta\alpha_{pb}$ is weak and corresponds to a pure pump-induced bleaching of the exciton resonance. This is what is to be expected, since by considering only the Pauli blocking contribution the exciton is approximated as a simple two-level system [39].

As shown by the dashed line in the lower part of Fig. 5a the first-order Coulomb contribution $\delta\alpha_{CI,1st}$ is stronger than $\delta\alpha_{pb}$ and antisymmetric around the exciton resonance. It has a dispersive line shape corresponding to a blue shift of the exciton, i.e., increased (decreased) absorption above (below) the resonance. As for the Pauli blocking this line shape of the first-order Coulomb term can be understood qualitatively well on the basis of a simple solution of the schematical equations of motion, i.e., (52) and (53) [39].

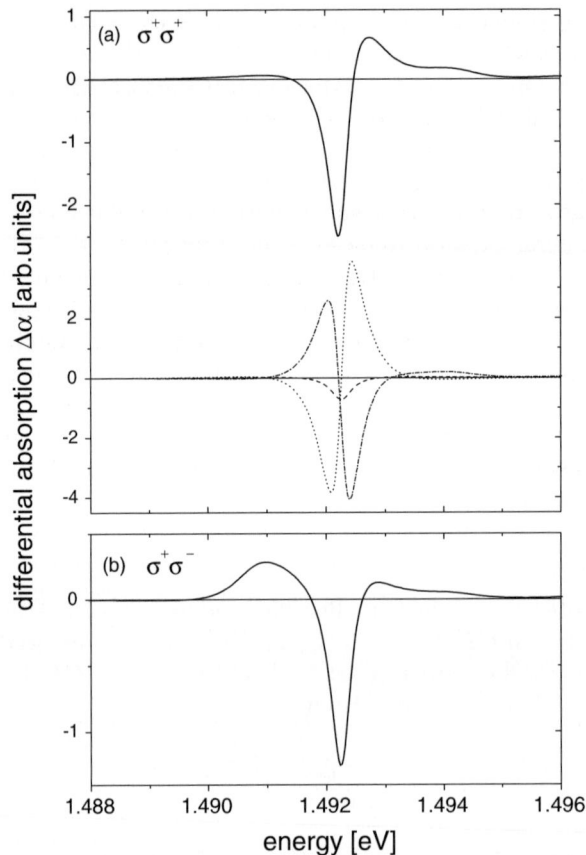

Fig. 5. Calculated differential absorption spectra for resonant excitation at the exciton. (**a**) co-circularly polarized pump- and probe-pulses ($\sigma^+\sigma^+$), the lower panel shows the three individual contributions to the signal (*solid*: Pauli blocking, *dashed*: first-order Coulomb, and *dotted*: Coulomb correlation). (**b**) opposite-circularly polarized pump- and probe-pulses ($\sigma^+\sigma^-$). Taken from [46]

The correlation contribution $\delta\alpha_{CI,corr}$ is shown as the dotted line in the lower part of Fig. 5a. Its line shape is again mainly dispersive around the exciton resonance, but with opposite sign compared to $\delta\alpha_{CI,1st}$, i.e., this term corresponds to a red shift. Besides nonvanishing values around the exciton energy, $\delta\alpha_{CI,corr}$ also includes terms which originate from resonances corresponding to transitions from excitons to unbound two-exciton states. These transitions are responsible for positive differential absorption, so called excited-state absorption, at energies above the exciton resonance.

When adding up the three contributions to obtain the total differential absorption, strong compensations occur between $\delta\alpha_{CI,1st}$ and $\delta\alpha_{CI,corr}$. The resulting differential absorption exhibits predominantly bleaching of the exciton resonance [39, 46].

We now discuss the results obtained when exciting with opposite circularly polarized pulses. For this polarization geometry, both, $\delta\alpha_{pb}$ and $\delta\alpha_{CI,1st}$, vanish as long as only transitions from the heavy-hole bands to the lowest electronic bands are considered and the system is spatially homogeneous [39]. This is due to the fact that none of these two contributions introduces any coupling between the subspaces of different spin states, since the Hartree–Fock approximation introduces no interaction among electrons and heavy-holes from the different spin-degenerate subbands. Therefore, for this polarization geometry, the differential absorption is given purely by the correlation contribution, i.e., $\delta\alpha = \delta\alpha_{CI,corr}$. As for co-circular excitation, also for opposite circular excitation, we find bleaching at the exciton resonance and excited-state absorption due to transitions to unbound two-exciton states which appears energetically above the exciton resonance, see Fig. 5b. Whereas for co-circularly polarized excitation only contributions from unbound two-exciton states are present, for opposite circular excitation also a clear signature of a bound biexciton is visible in the differential absorption spectrum, appearing about 1.4 meV below the excitonic resonance [39, 46]. As is shown in [46], the differential absorption spectra of Fig. 5 agree qualitatively with measurements performed on high-quality InGaAs quantum wells.

Numerical results for the excitonic optical Stark effect, i.e., the induced spectral changes around the exciton for detuned optical pumping below the resonance, are displayed in Fig. 6. To investigate the fundamental light-polarization dependence of the transient exciton shifts in semiconductors, we again concentrate on the case of circularly polarized pump and probe pulses. Figure 6a shows the computed differential absorption spectrum for co-circularly polarized pulses together with the separate contributions from the three different types of optical nonlinearities. Whereas the phase-space filling nonlinearity and the first-order Coulomb terms, i.e., the Hartree–Fock contributions induce a blue shift, the correlations alone would yield a red shift [46]. However, for the $\sigma^+\sigma^+$ excitation the magnitude of the correlation-induced contribution is relatively small and therefore the total signal is dominated by the blue shift of the Hartree–Fock terms.

The situation is, however, very different for the case of opposite circularly polarized $(\sigma^+\sigma^-)$ pump and probe pulses. As discussed above, for this configuration the Hartree–Fock contributions vanish. Therefore, the response is dominated by correlations involving heavy-hole excitons only.

As for $\sigma^+\sigma^+$ also for $\sigma^+\sigma^-$ excitation, Fig. 6b, the correlation part of the differential absorption represents a red shift. However, since the Hartree–Fock contributions vanish for this configuration, the correlation-induced red shift is not compensated by other terms and thus survives. This red shift is not directly related to the existence of a bound biexciton which can be excited with $\sigma^+\sigma^-$ polarized pulses. This can be seen already from the analysis of

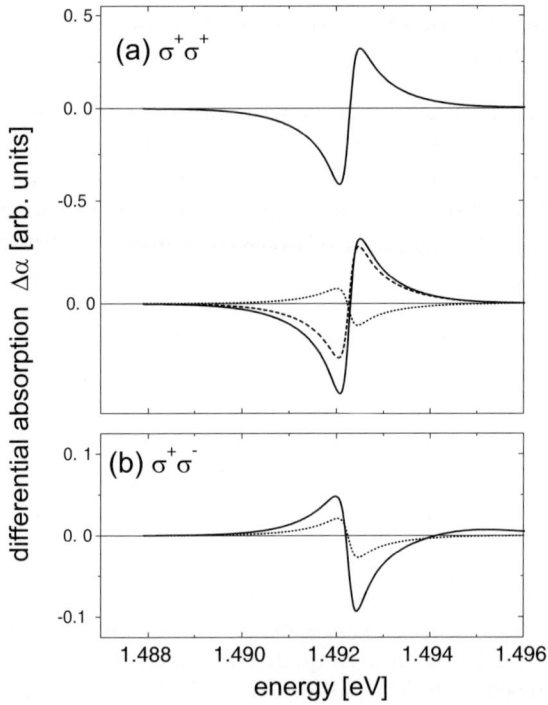

Fig. 6. Calculated differential absorption spectra for excitation 4.5 meV below the exciton and zero time delay. (a) co-circularly polarized pump- and probe-pulses ($\sigma^+\sigma^+$), the lower panel shows the three individual contributions to the signal (*solid*: Pauli blocking, *dashed*: first-order Coulomb, and *dotted*: Coulomb correlation). (b) opposite-circularly polarized pump- and probe-pulses ($\sigma^+\sigma^-$). The *dotted line* represents the result of a calculation within the second-order Born approximation including memory effects. After [46]

the $\sigma^+\sigma^+$ excitation configuration, where no bound biexciton is excited, but still the correlation term alone also amounts to a red shift of the exciton resonance.

Further insight is gained by additional calculations where the contributions from bound biexciton transitions for $\sigma^+\sigma^-$ excitation are artificially eliminated. This is done by performing calculations in the third-order limit of the second-order Born approximation including memory effects [46]. As shown in Fig. 6b, these calculations also produce a red shift where only the magnitude is somewhat reduced. If, however, the second-order Born approximation is combined with the Markov approximation, the red shift disappears for reasonable values of homogeneous broadening [3]. Therefore, one can conclude that the red shift is a genuine consequence of the dynamics, i.e., the memory character of the many-body Coulomb correlations.

4.5 Dynamics-Controlled Truncation Scheme: Coherent $\chi^{(5)}$-Limit

Here, we show schematically how the dynamics-controlled truncation scheme can be extended to the next nontrivial order, which is the coherent $\chi^{(5)}$-limit. To obtain the relevant equations in this order, one can apply the general scheme introduced in Sect. 4.3. In the following, we sketch very briefly how one has to extend the decoupling schemes to include the fifth-order contributions and give the resulting equations of motion schematically. More details can be found in [4, 18, 51, 52, 53].

Up to fifth order, the four-operator terms in the equation of motion for p can be expressed as

$$\langle a^\dagger a^\dagger b^\dagger a \rangle = p^*B + (B^* - p^*p^*)(pB - W) + O(E^7). \tag{57}$$

Here, a new quantity, i.e., the three-exciton transition $W \equiv \langle a^\dagger b^\dagger a^\dagger b^\dagger a^\dagger b^\dagger \rangle$, is introduced. Furthermore, also additional terms in the conservation laws for the electron and hole occupations, f^{ee} and f^{hh}, have to be considered. Up to fourth order one obtains for the electron occupation

$$\langle a^\dagger a \rangle = p^*p + (B^* - p^*p^*)(B - pp) + O(E^6), \tag{58}$$

and an identical schematical equation for $\langle b^\dagger b \rangle$.

In the coherent $\chi^{(5)}$-limit, we have to describe the dynamical evolution of single-, double-, and triple-electron-hole pair transitions, p, B, and W, respectively. Schematically the equations of motion for these quantities (without removing the lower-order correlations) read

$$i\hbar \frac{\partial}{\partial t} p = -\varepsilon_p\, p$$
$$+ \left[\, \mu\, -\, b\, (\, p^*p\, +\, (B^* - p^*p^*)\, (B - pp)\,)\, \right] E$$
$$+ V_{corr}\, [\, p^*B\, -\, (B^* - p^*p^*)\, (pB - W)\,], \tag{59}$$

$$i\hbar \frac{\partial}{\partial t} B = -\varepsilon_B\, B$$
$$+ (\, \tilde{\mu}\, p\, -\, \tilde{b}\, p^*B\,)\, E,$$
$$+ \tilde{V}_{corr}\, p^*W \tag{60}$$

$$i\hbar \frac{\partial}{\partial t} W = -\varepsilon_W\, W$$
$$+ \tilde{\tilde{\mu}}\, BE. \tag{61}$$

4.6 Signatures of Coherent Four-Particle Correlations Up to $\chi^{(5)}$

Whereas in the coherent $\chi^{(3)}$-limit the optical response of semiconductors is fairly well understood, not so much knowledge is available regarding the influence of many-exciton correlations beyond this low-intensity regime. This is partly due to the fact, that already in the coherent $\chi^{(5)}$-limit on top of the dynamics of one- and two-excitons, in principle, also three-exciton states W need to be considered. Microscopic calculations of the nonlinear optical response including three-exciton resonances represent solutions of a quantum-mechanical six-body problem. Such numerical evaluations of the resulting equations are currently only feasible for very small model systems, e.g., small and narrow semiconductor nanorings [54], but not for extended semiconductors. For systems like quantum wires and wells, to the best of our knowledge, calculations that fully include the dynamics of three exciton resonances have not been performed yet. It has, however, been possible to theoretically describe some experimental observations that arise with increasing intensity in four-wave mixing and pump probe by neglecting the dynamics of three-exciton states or by treating them approximately [52, 53, 55, 56, 57, 58].

In the following, we analyze the importance of four-particle correlations beyond the third-order limit, by studying the intensity dependence of the differential absorption considering pumping resonantly at and spectrally below the exciton. For this purpose, we neglect the three-exciton resonances W in our solutions of the microscopic versions of the coherent $\chi^{(5)}$-equations. The signals calculated numerically within this approximation are in good qualitative agreement with measurements performed on high-quality InGaAs quantum wells [52].

Numerical results for resonant excitation and 2 ps time delay between the pump and probe pulses are shown in Fig. 7. For these conditions the line shape of the pure $\chi^{(5)}$-contribution corresponds basically to the negative of the $\chi^{(3)}$ result. Consequently, as in the experiment, see [52], also in the calculations, the bleaching and the induced absorption decrease with increasing intensity in the normalized spectra. The numerical results also show a small shift of the bleaching maximum towards higher energies with increasing intensity, see Fig. 7, corresponding to a change in the line shape of the absorption peak. This shift of the bleaching maximum is induced by the induced absorption due to transitions to unbound two-exciton resonances. Figure. 7 furthermore demonstrates that the biexciton-induced polarization dependence of the optical response also remains present for elevated pump intensities.

Numerical results on the polarization and pump-intensity dependent absorption changes for off-resonant excitation are shown in Fig. 8. In agreement with the experiment, see [52], for co-circular excitation, we find exciton broadening in $\chi^{(5)}$. This broadening leads to distinct changes of the differential absorption line shape with increasing pump intensity. For this excitation configuration both, the $\chi^{(3)}$ and $\chi^{(5)}$ results, can be qualitatively understood already within the Hartree–Fock limit. Actually, describing the $1s$-exciton as a simple two-level system, which corresponds to neglecting all four- and six-particle correlations proportional to B and W in the coherent $\chi^{(5)}$-equations, is sufficient to reproduce both the shift in $\chi^{(3)}$ as well as the broadening in

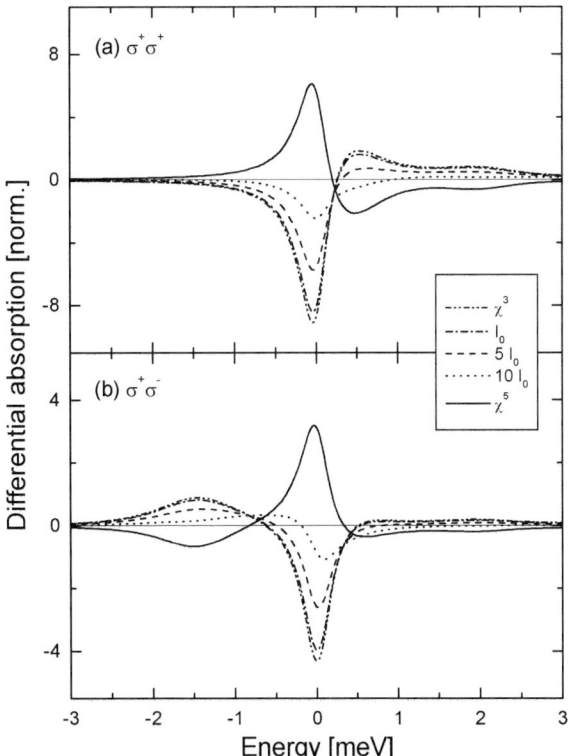

Fig. 7. Normalized calculated differential absorption spectra for excitation resonant at the exciton resonance and a delay of 2 ps between the pump and probe pulses for various pump pulse intensities. (**a**) co-circularly and (**b**) opposite-circularly polarized pump and probe pulses. Also the line shape of the calculated pure $\chi^{(5)}$-contribution is shown; its amplitude has been scaled. The zero of the energy scale coincides with the position of the exciton resonance in the linear absorption spectrum. Taken from [52]

$\chi^{(5)}$. This means that our results confirm those of [59], where it was shown that due to phase-space filling effects an off-resonant pump may induce exciton broadening type signatures in absorption if pump and probe pulses are overlapping in time.

For off-resonant and opposite-circularly polarized excitation, the correlation-induced red shift is decreasing in the normalized spectra with increasing pump intensity, see Fig. 8b. This is due to the fact that in this case the calculated pure $\chi^{(5)}$-contribution corresponds to a blue shift. Altogether, the experimentally measured polarization and intensity dependence of the excitonic optical Stark effect, including the differences in amplitude between co- and opposite-circular excitation [52], is very well reproduced by our model calculations. Even the ratio of the pump intensities necessary to

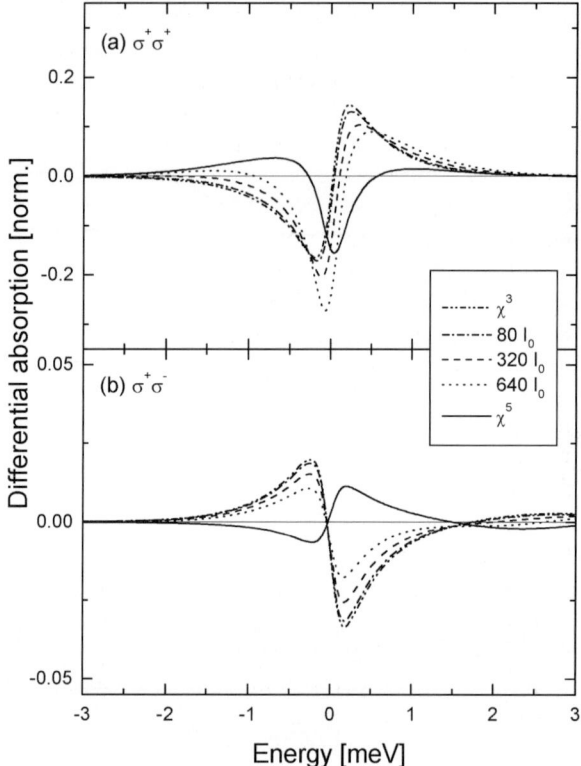

Fig. 8. Normalized calculated differential absorption spectra for excitation 4.5 meV below the exciton and zero delay between the pump and probe pulses for various pump pulse intensities. (**a**) co-circularly and (**b**) opposite-circularly polarized pump and probe pulses. Also the line shape of the calculated pure $\chi^{(5)}$-contribution is shown; its amplitude has been scaled. The zero of the energy scale coincides with the position of the exciton resonance in the linear absorption spectrum. Taken from [52]

observe $\chi^{(5)}$ effects for resonant and nonresonant excitation agrees quite well with experiment [52].

5 Conclusions and Outlook

We have summarized here the semiclassical approach of a microscopic theory for the nonlinear optical response of semiconductors, where the light is treated as a classical electromagnetic field and the optical polarization is computed quantum mechanically. This theory leads to coupled equations for the microscopic interband polarization and the carrier occupation functions. At the Hartree–Fock level, the resulting semiconductor Bloch equations contain many-body contributions via bandgap and field renormalization.

Two approaches have been described which are capable of analyzing many-body Coulomb correlations beyond the time-dependent Hartree–Fock approximation. The second-order Born approximation includes scattering among occupations and polarizations as well as the screening of the Coulomb interaction potential. This approach is nonperturbative in the optical field and thus can be applied to analyze the optical response when exciting with high intensity pulses and in the presence of a high-density electron-hole plasma. The dynamics-controlled truncation scheme, on the other hand, is based on an expansion of the nonlinear optical response in powers of the optical field. Unlike the second-order Born approximation, this scheme includes explicitly the dynamics of biexcitonic four-particle correlations. It is therefore well suited to describe signatures of these four-particle correlations in the coherent nonlinear optical response for excitation with moderate intensities.

The examples mentioned in this review are just a few selected cases where Coulomb many-body correlation effects have been studied at the level of a microscopic theory. More examples can be found in the cited references. As is shown there, the predictions of the microscopic theory are often in good agreement with experiments performed on semiconductor nanostructures.

Despite the advanced level of understanding of light-matter interaction effects in semiconductors, there are still many open questions and challenges. One problem of recent interest concerns the generation dynamics of excitonic populations, see, e.g., [28, 41] and references therein. Under which conditions do excitons form and can one reliably identify excitonic condensation effects in systems where the Fermionic nature of the constituent electrons and holes cannot be neglected?

Many open problems are related to the analysis of interacting carrier systems with structural disorder. Basically all realistic, i.e., experimentally realizable systems that are structured on a mesoscopic scale, like quantum wells, wires, or dots, suffer from disorder due to compositional and interface fluctuations. There is no translational invariance in such systems such that a separation of relative- and center-of-mass motion is not possible. Therefore, in this case not even the linear excitonic absorption problem can be solved analytically. Even though there is some, mostly numerical work, see, e.g., [40], which investigates the disorder-induced dephasing of the four-wave-mixing signal, the field of interactions and disorder on mesoscopic scales is still wide open.

In the area of quantum optical effects in semiconductors, the existing work has also barely scratched the surface. Relatively few measurements of true quantum signatures in semiconductors have been reported, see, e.g., [60, 61, 62]. One of the big challenges is to find ways to reach the truly quantum statistical limit. This would be a big step not only on our way of understanding the fundamentals of light-matter interaction, but also in the direction of potential quantum logic applications of semiconductor structures.

Acknowledgments

We have benefited greatly from many discussions and collaborations with H.M. Gibbs, M. Kira, and P. Thomas and their research groups. This work is supported by the Deutsche Forschungsgemeinschaft (DFG), by the Max-Planck Research prize of the Humboldt and Max-Planck societies, and by the Center for Optodynamics, Philipps University, Marburg, Germany. SWK acknowledges partial support through AFOSR MRI F46920-02-1-0380. TM thanks the DFG for support via a Heisenberg fellowship (ME 1916/1). We thank the John von Neumann Institut für Computing (NIC), Forschungszentrum Jülich, Germany, for grants for extended CPU time on their supercomputer systems.

References

1. H. Haug and S.W. Koch, *Quantum Theory of the Optical and Electronic Properties of Semiconductors*, 4th ed., World Scientific, Singapore (2004)
2. S.W. Koch, M. Kira, and T. Meier, invited review, J. of Optics B **3**, R29 (2001)
3. T. Meier and S.W. Koch, in *Ultrafast Physical Processes in Semiconductors*, ed. K.T. Tsen, Semiconductors and Semimetals Vol. 67, Academic Press (2001), p. 231
4. T. Meier, C. Sieh, M. Kira, S.W. Koch, Y.-S. Lee, T.B. Norris, F. Jahnke, G. Khitrova und H.M. Gibbs, in *Optical Microcavities*, ed. K. Vahala, World Scientific, Singapore (2004), p. 239
5. R.J. Elliott, in *Polarons and Excitons*, eds. C.G. Kuper and G.D. Whitefield, Oliver and Boyd, (1963), p. 269
6. H. Haug and S. Schmitt-Rink, Prog. Quant. Electron. **9**, 3 (1984)
7. M. Lindberg and S.W. Koch, Phys. Rev. B **38**, 3342 (1988)
8. M. Lindberg, R. Binder, and S.W. Koch, Phys. Rev. A **45**, 1865 (1992)
9. M. Lindberg, Y.Z. Hu, R. Binder, and S.W. Koch, Phys. Rev. B **50**, 18060 (1994)
10. V.M. Axt and A. Stahl, Z. Phys. B **93**, 195 (1994); *ibid.*, 205 (1994)
11. R. Binder and S.W. Koch, Prog. Quant. Electron. **19**, 307 (1995)
12. T. Östreich, K. Schönhammer, and L.J. Sham, Phys. Rev. Lett. **74**, 4698 (1995)
13. W. Schäfer, Journ. Opt. Soc. Am. B **13**, 1291 (1996)
14. F. Jahnke, M. Kira, and S.W. Koch, Z. Physik B **104**, 559 (1997)
15. S.W. Koch, T. Meier, F. Jahnke, and P. Thomas, Appl. Phys. A **71**, 511 (2000)
16. N.H. Kwong and R. Binder, Phys. Rev. B **61**, 8341 (2000)
17. F. Rossi and T. Kuhn, Rev. Mod. Phys. **74**, 895 (2002)
18. W. Schäfer and M. Wegener, *Semiconductor Optics and Transport Phenomena*, Springer, Berlin (2002)
19. E.P. Wigner, Phys. Rev. **46**, 1002 (1934)
20. S. Haas, F. Rossi, and T. Kuhn, Phys. Rev. B **53**, 12855 (1996)
21. G. Khitrova, H.M. Gibbs, F. Jahnke, M. Kira, and S.W. Koch, Rev. Mod. Phys. **71**, 1591 (1999)
22. N. Bloembergen, *Nonlinear Optics*, Benjamin Inc., New York (1965)

23. D.S. Chemla and J. Shah, Nature **411**, 549 (2001)
24. M. Kira, F. Jahnke, and S.W. Koch, Phys. Rev. Lett. **81**, 3263 (1998)
25. M. Kira, W. Hoyer, F. Jahnke, and S.W. Koch, Prog. Quantum Electron. **23**, 189 (1999)
26. M. Kira, F. Jahnke, and S.W. Koch, Phys. Rev. Lett. **82**, 3544 (1999)
27. M. Kira, W. Hoyer, T. Stroucken, and S.W. Koch, Phys. Rev. Lett. **87**, 176401 (2001)
28. W. Hoyer, M. Kira, and S.W. Koch, Phys. Rev. B **67**, 155113 (2003)
29. W. Schäfer, R. Lövenich, N.A. Fromer, and D.S. Chemla, Phys. Rev. Lett. **86**, 344 (2001)
30. R. Lövenich, C.W. Lai, D. Hägele, D.S. Chemla, and W. Schäfer, Phys. Rev. B **66**, 045306 (2002)
31. W. Schäfer, D.S. Kim, J. Shah, T.C. Damen, J.E. Cunningham, K.W. Goosen, L.N. Pfeiffer, and K. Köhler, Phys. Rev. B **53**, 16429 (1996)
32. K. Bott, O. Heller, D. Bennhardt, S.T. Cundiff, P. Thomas, E.J. Mayer, G.O. Smith, R. Eccleston, J. Kuhl, and K. Ploog, Phys. Rev. B **48**, 17418 (1993)
33. S.W. Koch, T. Meier, W. Hoyer, and M. Kira, invited review, Physica E **14**, 45 (2002)
34. F. Jahnke, M. Kira, S.W. Koch, G. Khitrova, E.K. Lindmark, T.R. Nelson, D.V. Wieck, J.D. Berger, O. Lyngnes, H.M. Gibbs, and K. Tai, Phys. Rev. Lett. **77**, 5257 (1996)
35. H. Wang, K.B. Ferrio, D.G. Steel, Y.Z. Hu, R. Binder, and S.W. Koch, Phys. Rev. Lett. **71**, 1261 (1993)
36. T. Rappen, U.G. Peter, M. Wegener, and W. Schäfer, Phys. Rev. B **49**, 7817 (1994)
37. O. Lyngnes, J.D. Berger, J.P. Prineas, S. Park, G. Khitrova, H.M. Gibbs, F. Jahnke, M. Kira, and S.W. Koch, Solid State Commun. **104**, 297 (1997)
38. W.W. Chow and S.W. Koch, *Semiconductor-Laser Fundamentals*, Springer, Berlin (1999)
39. C. Sieh, T. Meier, A. Knorr, F. Jahnke, P. Thomas, and S.W. Koch, Europ. Phys. Journ. B **11**, 407 (1999)
40. S. Weiser, T. Meier, J. Möbius, A. Euteneuer, E.J. Mayer, W. Stolz, M. Hofmann, W.W. Rühle, P. Thomas, and S.W. Koch, Phys. Rev. B **61**, 13088 (2000)
41. S. Chatterjee, C. Ell, S. Mosor, G. Khitrova, H.M. Gibbs, W. Hoyer, M. Kira, S.W. Koch, J.P. Prineas, and H. Stolz, Phys. Rev. Lett. **92**, 067402 (2004)
42. V.M. Axt, A. Stahl, E.J. Mayer, P. Haring Bolivar, S. Nüsse, K. Ploog, and K. Köhler, phys. stat. sol. (b) **188**, 447 (1995)
43. P. Kner, S. Bar-Ad, M.V. Marquezini, D.S. Chemla, and W. Schäfer, Phys. Rev. Lett. **78**, 1319 (1997)
44. T. Östreich, K. Schönhammer, and L. J. Sham, Phys. Rev. B **58**, 12920 (1998)
45. P. Kner, W. Schäfer, R. Lövenich, and D.S. Chemla, Phys. Rev. Lett. **81**, 5386 (1998)
46. C. Sieh, T. Meier, F. Jahnke, A. Knorr, S.W. Koch, P. Brick, M. Hübner, C. Ell, J. Prineas, G. Khitrova, and H.M. Gibbs, Phys. Rev. Lett. **82**, 3112 (1999)
47. S.W. Koch, C. Sieh, T. Meier, F. Jahnke, A. Knorr, P. Brick, M. Hübner, C. Ell, J. Prineas, G. Khitrova, and H.M. Gibbs, J. of Lumin. **83/84**, 1 (1999)
48. T. Meier, S.W. Koch, M. Phillips, and H. Wang, Phys. Rev. B **62**, 12605 (2000)

49. M.E. Donovan, A. Schülzgen, J. Lee, P.-A. Blanche, N. Peyghambarian, G. Khitrova, H.M. Gibbs, I. Rumyantsev, N.H. Kwong, R. Takayama, Z.S. Yang, and R. Binder, Phys. Rev. Lett. **87**, 237402 (2001)
50. P. Brick, C. Ell, G. Khitrova, H.M. Gibbs, T. Meier, C. Sieh, and S.W. Koch, Phys. Rev. B **64**, 075323 (2001)
51. K. Victor, V.M. Axt, and A. Stahl, Phys. Rev. B **51**, 14164 (1995)
52. T. Meier, S.W. Koch, P. Brick, C. Ell, G. Khitrova, and H.M. Gibbs, Phys. Rev. B **62**, 4218 (2000)
53. W. Langbein, T. Meier, S.W. Koch, and J.M. Hvam, J. Opt. Soc. Am. B **18**, 1318 (2001)
54. T. Meier, C. Sieh, E. Finger, W. Stolz, W.W. Rühle, P. Thomas, and S.W. Koch, phys. stat. sol. (b) **238**, 537 (2003)
55. G. Bartels, V.M. Axt, K. Victor, A. Stahl, P. Leisching, and K. Köhler, Phys. Rev. B **51**, 11217 (1995)
56. T.F. Albrecht, K. Bott, T. Meier, A. Schulze, M. Koch, S.T. Cundiff, J. Feldmann, W. Stolz, P. Thomas, S.W. Koch, and E.O. Göbel, Phys. Rev. B **54**, 4436 (1996)
57. S.R. Bolton, U. Neukirch, L.J. Sham, D.S. Chemla, and V.M. Axt, Phys. Rev. Lett. **85**, 2002 (2000)
58. H.P. Wagner, H.-P. Tranitz, M. Reichelt, T. Meier, and S.W. Koch, Phys. Rev. B **64**, 233303 (2001)
59. R. Binder, S.W. Koch, M. Lindberg, W. Schäfer, and F. Jahnke, Phys. Rev. B **43**, 6520 (1991)
60. Y.S. Lee, T.B. Norris, M. Kira, F. Jahnke, S.W. Koch, G. Khitrova, and H.M. Gibbs, Phys. Rev. Lett. **83**, 5338 (1999)
61. P. Michler, A. Kiraz, C. Becher, W.V. Schoenfeld, P.M. Petroff, L.D. Zhang, E. Hu, and A. Imamoglu, Science **290**, 2282 (2000)
62. C. Ell, P. Brick, M. Hübner, E.S. Lee, O. Lyngnes, J.P. Prineas, G. Khitrova, H.M. Gibbs, M. Kira, F. Jahnke, S.W. Koch, D.G. Deppe, and D.L. Huffaker Phys. Rev. Lett. **85**, 5392 (2000)

Exciton and Polariton Condensation

D. Porras[1], J. Fernandez-Rossier[2], and C. Tejedor[3]

[1] Max-Planck Institute for Quantum Optics, 85748, Garching, Germany
[2] Departamento de Física Aplicada, Universidad de Alicante, 03080 Alicante, Spain
[3] Departamento de Física Teórica de la Materia Condensada, Universidad Autónoma de Madrid, 28049 Madrid, Spain

Abstract. These are the notes for a course on a theoretical description of condensation of excitons and exciton-polaritons in semiconductors. A special emphasis is made on the case of quantum wells. We start by presenting a standard theory that can be found also in several excellent books and reviews. We concentrate in the question of detecting condensation without paying attention to the *open* problem of the dynamics of condensation appearance. Special care is devoted to the emission of light. The recently studied case of condensation of magnetoexcitons is briefly discussed. Since excitons have the third component of the total angular momentum as an internal degree of freedom, this opens the extremely interesting possibility of multicomponent condensates which we discuss in some detail. In particular, we show the appearance of an interesting behavior of the polarization of the light emitted by this multicomponent condensate.

An important problem for exciton condensation is the life-time of these quasi-particles as well as their large effective masses which imply a large threshold density for condensation. A possible alternative to avoid these difficulties is to work with exciton-polaritons in quantum wells embedded in microcavities. We describe carefully the stimulated scattering processes that can produce a large density of these quasi-particles in a given state as a necessary ingredient for condensation. Once such a large population is obtained, the corresponding polariton laser shows special characteristics in its spectrum that we describe in the final part of these notes.

1 Introduction

These are notes for a course on some theoretical aspects of *exciton and polariton condensation*. They are not exhaustive on all the aspects in the field. In particular, we do not include a description on the experimental situation which is currently developing very fast [1]. Excellent descriptions of many topics in this field are contained in reference [2, 3].

The topics covered here are:

1. Exciton condensation: Standard theory
2. Emission of light
3. Magnetoexcitons
4. Multicomponent condensates
5. Polariton condensation
6. Polariton laser

D. Porras et al.: *Exciton and Polariton Condensation*, Lect. Notes Phys. **689**, 153–189 (2006)
www.springerlink.com

An exciton is the lowest electronic excitation of a semiconductor. It corresponds to the creation of an electron-hole pair with the electron lying in the semiconductor conduction band while the hole is the lack of occupation of a valence band state. When the semiconductor is illuminated with moderate intensity, a large set of electron-hole pairs is formed and one can make a description in terms on an exciton gas. If the optical excitation is very intense, the density is so high that a two component plasma of electrons and holes is a better picture of the state of the system. Here we will concentrate in the moderate excitation regime and study the exciton gas.

The excitation energy of the exciton as a quasi-particle is the semiconductor band gap minus a binding energy which is essentially determined by the Coulomb attraction between the electron and the hole forming the pair with wave function $\varphi(\mathbf{r})$ where $\mathbf{r}$ is the electron-hole relative coordinate. The exciton binding energy is typically several meV which implies a lifetime for this quasiparticle in the range of nanoseconds. States at the conduction band have angular momentum equal to $1/2$ while states at the valence have either angular momentum $3/2$ (heavy holes) or $1/2$ (light holes). Therefore, exciton can have angular momentum -2, -1 0, 1 or 2. Only excitons with angular momentum ± 1 can be excited by dipolar optical excitation so that they are usually denoted as bright excitons, while excitons with angular momentum ± 2 or 0 are labeled as dark excitons. Although dark excitons can play an important an important role in the dynamics of the excitation of the system, we will concentrate here on bright excitons because we will pay special attention to the optical properties of the system.

Condensation of excitons can occur in two physically different ways: the excitation of a semiconductor by coherent light can produce an exciton gas with a coherence induced by that of the excitation source. This is the origin of interesting non-linear optical properties which we are not going to cover here [4, 5]. We will concentrate in the second alternative in which the system is non-resonantly excited. Excitons relax to thermal equilibrium and a spontaneous symmetry breaking produces macroscopic coherence of the system. Obviously, this process requires relaxation time to be much shorter than exciton lifetime, a property which is experimentally accessible.

Let us start by clarifying the meaning of words as coherence and condensation.

1.1 Coherence

We are going to work with bosonic fields, i.e. with harmonic oscillators. In order to have a description of a quantum harmonic oscillator as close as possible to a classical harmonic oscillator, one builds up wave packets with a minimum uncertainty $\Delta P \Delta Q = 1/2$ (we take $\hbar = 1$ except when we give explicit expressions for transitions temperatures) for its generalized coordinate Q and momentum P. Since the minimum uncertainty condition can be rewritten [6] as $\Delta P^2 \Delta Q^2 = 1/4$, it is usual to introduce a squeezing parameter r such that, $\Delta P^2 = e^{-2r}/2$ and $\Delta Q^2 = e^{2r}/2$. When $r = 0$, the wave packet is labeled as coherent state having the highest analogy with a classical state because its shape is time-independent. When $r \neq 0$, the shape varies with time and one has squeezed states (either in P or in Q) which do not have any classical analog, being only possible as quantum mechanical states.

In our case, the coordinates and momenta of interest are the occupation n and the phase ϕ (although the phase operator has problems to be precisely defined). Coherent states are eigenstates of the creation operator

$$a \mid \alpha \rangle = \alpha \mid \alpha \rangle \tag{1}$$

with $\alpha = \mid \alpha \mid e^{i\phi}$. This property implies a great advantage: the coherent state describes a system with a non-zero order parameter $\langle a \rangle = \alpha \neq 0$. In terms of Fock states,

$$\mid \alpha \rangle = \sum_n \mid n \rangle \langle n \mid \alpha \rangle = \sum_n \mid n \rangle \langle 0 \mid \frac{a^n}{\sqrt{n!}} \mid \alpha \rangle = \langle 0 \mid \alpha \rangle \sum_n \frac{\alpha^n}{\sqrt{n!}} \mid n \rangle . \tag{2}$$

A coherent state has a Poisson distribution in terms of Fock states, but the opposite is not always true: for instance, a Poisson statistical mixture of Fock states is not a coherent state. Once the state is normalized, one gets

$$\mid \alpha \rangle = e^{-|\alpha|^2/2} \sum_{n=0}^{\infty} \frac{\alpha^n}{\sqrt{n!}} \mid n \rangle = e^{-|\alpha|^2/2} \sum_{n=0}^{\infty} \frac{\alpha^n a^{\dagger n}}{n!} \mid 0 \rangle = e^{-|\alpha|^2/2} e^{\alpha a^\dagger} \mid 0 \rangle . \tag{3}$$

Since $e^{-\alpha^* a} \mid 0 \rangle = \mid 0 \rangle$, any coherent state is a displaced ground state of the oscillator

$$\mid \alpha \rangle = e^{-|\alpha|^2/2} e^{\alpha a^\dagger} e^{-\alpha^* a} \mid 0 \rangle = D(\alpha) \mid 0 \rangle . \tag{4}$$

The oscillator displacement involved in a coherent state implies a breaking of the Hamiltonian symmetry. As mentioned above when introducing the coherent state, a finite order parameter $\langle \alpha \rangle$ characterizes this symmetry breaking. The mean number of bosons in a coherent state is $N = \langle n \rangle = \mid \alpha \mid^2$. When N is large, one gets uncertainties $\Delta N = \sqrt{N}/2$ and $\Delta \phi = 1/(\sqrt{N})$. Moreover, N large implies a macroscopic displacement $\alpha \propto \sqrt{N} e^{i\phi} \propto \sqrt{\mathcal{V}}$ where $\mathcal{V}$ is the volume of the system. Therefore, one says that the coherent state is macroscopic. Bosons are in states $\mathbf{k}$ which are plane waves in the homogeneous case. Macroscopic occupation of a given state $\mathbf{k}_0$ giving a coherent state implies the existence of off-diagonal long range order (ODLRO). Introducing a local Bose field

$$a(\mathbf{r}) = \frac{1}{\sqrt{\mathcal{V}}} \sum_{\mathbf{k}} a_{\mathbf{k}} e^{i\mathbf{k}\mathbf{r}}, \tag{5}$$

the off-diagonal correlation

$$\langle a^\dagger(\mathbf{r}')a(\mathbf{r})\rangle = \frac{N_{\mathbf{k}_0}}{\mathcal{V}}e^{i\mathbf{k}_0(\mathbf{r}'-\mathbf{r})} + \frac{1}{\mathcal{V}}\sum_{\mathbf{k}\neq\mathbf{k}_0}\langle a_{\mathbf{k}}^\dagger a_{\mathbf{k}}\rangle e^{i\mathbf{k}(\mathbf{r}'-\mathbf{r})} \tag{6}$$

is finite for $|\mathbf{r}'-\mathbf{r}|\to\infty$ due to the first term. ODLRO is a signature of the coherence of a macroscopic state.

1.2 Condensation

Condensation is a phase transition in which a large number of particles become in a macroscopic coherent state. This collective effect does not affect to all the particles in the system. In fact, since the state is coherent, the number of particles is not completely well defined but, N is much larger than its uncertainty $\sqrt{N}$. The condensation takes place at a critical temperature T_c which has different expressions for different kinds of condensations. Condensation occurs for non interacting gases, e.g. Bose–Einstein condensation (BEC), for strongly interacting systems, e.g. superconductivity in which attraction creates pairs which condense and for intermediate situations. Condensation occurs for bosons, e.g. BEC of bosonic atoms and ^{4}He superfluid, and fermions either in pairs, e.g. superconductivity, or unpaired, e.g ^{3}He and QHE. The case of excitons has the complication of being quasiparticles with a rather bosonic character for low densities while the fermioninc character of electrons and holes becomes essential when density increases.

1.3 Bosonic Limit of Excitons

The lowest energy excitation of a semiconductor is the promotion of an electron from the valence to the conduction band. This is a quasiparticle, called exciton, created by an operator (we take $m_e = m_h$ to simplify the equations)

$$\Psi_{\mathbf{k}}^\dagger = \frac{1}{\sqrt{2\mathcal{V}}}\sum_{\mathbf{q},\sigma}\varphi(\mathbf{k}/2+\mathbf{q})c_{\mathbf{k}+\mathbf{q},\sigma}^\dagger d_{-\mathbf{q},-\sigma}^\dagger \tag{7}$$

where $\varphi(\mathbf{q})$ is the exciton wave function, $\mathbf{k}$ and $\mathbf{q}$ wave vectors, σ the spin and $c^\dagger$ and $d^\dagger$ create electrons and holes in the conduction and valence bands respectively. The algebra of excitonic operators is

$$\left[\Psi_{\mathbf{k}},\Psi_{\mathbf{k}'}^\dagger\right] = \delta_{\mathbf{k},\mathbf{k}'} - \frac{1}{2\mathcal{V}}\sum_{\mathbf{q},\sigma}\varphi^*(\mathbf{q}-\mathbf{k}'/2)\varphi(\mathbf{q}-\mathbf{k}/2)d_{\mathbf{k}-\mathbf{q},-\sigma}^\dagger d_{\mathbf{k}'-\mathbf{q},-\sigma}$$

$$-\frac{1}{2\mathcal{V}}\sum_{\mathbf{q},\sigma}\varphi^*(\mathbf{k}'/2-\mathbf{q})\varphi(\mathbf{k}/2-\mathbf{q})c_{\mathbf{k}-\mathbf{q},\sigma}^\dagger c_{\mathbf{k}'-\mathbf{q},\sigma}. \tag{8}$$

This commutator is the one corresponding to bosons (the first term) plus a correction linear on the overlap between the two excitonic wave functions. In other words, excitons behave as bosons in the dilute limit while the fermionic character of electrons and holes becomes more and more important when density increases. The parameter quantifying these regions is $n_X(a_B)^D$, where n_X is the density of excitons, a_B the exciton Bohr radius and D the dimensionality of the system. For practical purposes, one can consider the excitons as bosons for $n_X a_B^D < 0.1$. Although the physics behind condensation can be different among different regimes, Keldysh and collaborators [2] show that a common treatment is possible in any regime.

In some systems, the lowest energy state is not a certain density of excitons but a set of biexcitons (exciton pairing to lower the energy). The condensation of biexcitons can also occur in a way similar to that of excitons so that we will not pay special attention to such a case.

2 Exciton Condensation: Standard Theory

Let us go, step by step, toward exciton condensation.

2.1 Non-Interacting Bosons

BEC occurs at the critical temperature in which BE distribution can not accommodate, with a non-positive chemical potential μ, all the particles in the different states of the spectrum. Therefore a finite density of particles n_0 must occupy the lowest energy state $\mathbf{k}_0$. This means that $\langle a_0^\dagger a_0 \rangle = n_0 \mathcal{V}$, so that, in the thermodynamic limit, the operators $a_0^\dagger$ and a_0 can be substituted by c-numbers

$$a_0^\dagger = \sqrt{\mathcal{V}n_0}e^{i\phi} \quad ; \quad a_0 = \sqrt{\mathcal{V}n_0}e^{-i\phi} \ . \tag{9}$$

This is not compatible with the fact that the Hamiltonian of non-interacting bosons commutes with any number operator, in particular with $a_0^\dagger a_0$ implying the conservation of the number of particles in the lowest energy state. Particle conservation in the lowest energy state has the mathematic expression $\langle a_0 \rangle = \langle a_0^\dagger \rangle = 0$ which is not compatible with finite values given above for the thermodynamic limit.

The way in which the system can accommodate a finite density n_0 in the lowest energy state is by spontaneously breaking the gauge symmetry $[H, a_0^\dagger a_0] = 0$. As discussed for coherent states, this is made by means of a displacement $D(\alpha)$ with $\alpha = \sqrt{\mathcal{V}n_0}e^{i\phi}$ so that the ground state becomes the coherent state $\mid \alpha \rangle$ involving an order parameter $\langle a_0 \rangle = \sqrt{\mathcal{V}n_0}e^{-i\phi}$.

The critical temperature for BEC is obtained by the condition $\mu = 0$. This gives $T_c = 0$ in 2D while in 3D

$$T_c = \frac{2\pi\hbar^2}{1.897 m k_B} \left(\frac{n}{\mathcal{D}_{spin}} \right)^{2/3} , \tag{10}$$

where m and n are the mass and total density of bosons and $\mathcal{D}_{spin}$ the spin degeneracy. Typical parameters for excitons, give critical temperatures of a few Kelvins, i.e. much higher than for BEC of atoms. BEC at finite temperature is possible in 2D either by imperfections or finite size of the system [7]. For instance, an harmonic trap, of characteristic frequency ω_0, in 2D implies a critical BEC temperature $T_c = \hbar\omega_0\sqrt{6N_T}/\pi k_B\sqrt{\mathcal{D}_{spin}}$, where N_T is the number of bosons in the trap.

The condensate of non-interacting bosons is unstable with respect spatial variations of the phase ϕ [7]. This problem disappears with a weak interaction as the one we are going to consider now.

2.2 Weakly Interacting Bosons

We are going to describe the well known Bogoliubov theory for superfluids. Such a theory involves a perturbation procedure which is not very good for ^4He because there the interaction is too strong. However, it is good a candidate to understand the case of low density excitons where the interaction is much weaker. The interaction $U(\mathbf{r})$ between excitons depends on microscopic details of the semiconductor, but in general is short range and we will take it repulsive (if attractive, it will produce biexcitons that we do not consider here). The Hamiltonian describing excitons, with dispersion relation $\epsilon_X(\mathbf{k}) = \epsilon_X(0) + k^2/(2m_X)$, treated as bosons is

$$H = \sum_{\mathbf{k}}[\epsilon_X(\mathbf{k}) - \mu_X]a_{\mathbf{k}}^{\dagger}a_{\mathbf{k}} + \frac{1}{2V}\sum_{\mathbf{k},\mathbf{k}',\mathbf{q}} U_{\mathbf{q}}a_{\mathbf{k}}^{\dagger}a_{\mathbf{k}'}^{\dagger}a_{\mathbf{k}'+\mathbf{q}}a_{\mathbf{k}-\mathbf{q}} \ . \tag{11}$$

Symmetry breaking is described, as above, by displacing the $\mathbf{k}_0$ term of this Hamiltonian. This is equivalent to a transformation to the operators

$$\alpha_0 = a_0 - \sqrt{Vn_0} \ ; \quad \alpha_{\mathbf{k}} = a_{\mathbf{k}}(\mathbf{k} \neq \mathbf{k}_0) \ . \tag{12}$$

This transformation introduces a term in $\alpha_0^{\dagger}+\alpha_0$, but this term cancels out at the condensation where the chemical potential μ_X coincides with the lowest energy $\epsilon_X(0) + U_{\mathbf{k}_0}n_0$. Since Vn_0 is a macroscopically large number, all the terms having only $\mathbf{k} \neq \mathbf{k}_0$ can be treated perturbatively. This leaves a much simpler Hamiltonian which can be diagonalized by means of a Bogoliubov transformation

$$\tilde{\alpha}_{\mathbf{k}} = u_{\mathbf{k}}\alpha_{\mathbf{k}+\mathbf{k}_0} + v_{\mathbf{k}}\alpha_{\mathbf{k}_0-\mathbf{k}}^{\dagger} \tag{13}$$

with

$$u_{\mathbf{k}} = \sqrt{\frac{k^2/(2m_X) + U_{\mathbf{k}}n_{\mathbf{k}}}{2E(\mathbf{k})} + \frac{1}{2}} \ ; \quad v_{\mathbf{k}} = \sqrt{\frac{k^2/(2m_X) + U_{\mathbf{k}}n_{\mathbf{k}}}{2E(\mathbf{k})} - \frac{1}{2}} \tag{14}$$

so that $u_{\mathbf{k}}^2 - v_{\mathbf{k}}^2 = 1$. The total dispersion relation

$$-\frac{\mathcal{V}n_0^2 U_{\mathbf{k}_0}}{2} + E(\mathbf{k}) \tag{15}$$

has a gapless excitation part

$$E(\mathbf{k}) = \sqrt{[k^2/(2m_X)]^2 + 2U_{\mathbf{k}}n_{\mathbf{k}}k^2/(2m_X)}\,, \tag{16}$$

linear in k for long wavelengths.

The difficulty of getting clear insight on the physics behind condensation is clearly and carefully discussed in [2]: when Hanamura and Haug extended to finite temperature the Bogoliubov theory here described, they concluded that condensation is a second order phase transition. However, Lee and Yang made a different analysis to conclude that the transition is a first order one.

In 2D, a weak interaction does not allow a strict BEC with ODLRO. However, a spontaneous phase correlation, known as Kosterlitz-Thouless (KT) transition, can appear presenting off-diagonal correlation decreasing as a power law of the distance (instead of doing it exponentially). For an interaction characterized by a scattering length a, the KT critical temperature is not zero anymore, but instead $T_c = 4\pi\hbar^2 n/2mlnln(1/na^2)$.

2.3 Interacting Electron-Hole Pairs: Excitonic Insulator

In the previous subsection, we have seen how interaction effects between excitons play an important role. But this interaction is short range so that it works only when excitons are close to each other (even for a dilute gas). However, at short distances, the fermionic character of the two exciton components also play a role. Therefore, one can expect that even at low densities, considering actual electron-hole pairs is as important as considering interactions. Let us follow the descriptions of a condensate, at zero temperature, in terms of electron-hole pairs, as introduced by Keldysh and coworkers and Nozieres and coworkers [2, 8, 9, 10]. Neglecting electron-hole exchange, the Hamiltonian is

$$H_X = \sum_{\mathbf{k},\sigma}[\epsilon_e(\mathbf{k}) - \mu_{e,\sigma}]c_{\mathbf{k},\sigma}^\dagger c_{\mathbf{k},\sigma}$$

$$+ \sum_{\mathbf{k},\varsigma}[\epsilon_h(\mathbf{k}) - \mu_{h,\varsigma}]d_{\mathbf{k},\varsigma}^\dagger d_{\mathbf{k},\varsigma} + \sum_{\mathbf{k},\mathbf{k}',\mathbf{q},\sigma_1,\sigma_2,\varsigma_1,\varsigma_2}\frac{V_{\mathbf{q}}}{2}$$

$$\left[c_{\mathbf{k},\sigma_1}^\dagger c_{\mathbf{k}',\sigma_2}^\dagger c_{\mathbf{k}'+\mathbf{q},\sigma_2}c_{\mathbf{k}-\mathbf{q},\sigma_1} + d_{\mathbf{k},\varsigma_1}^\dagger d_{\mathbf{k}',\varsigma_2}^\dagger d_{\mathbf{k}'+\mathbf{q},\varsigma_2}d_{\mathbf{k}-\mathbf{q},\varsigma_1}\right.$$

$$\left.-2c_{\mathbf{k},\sigma_1}^\dagger d_{\mathbf{k}',\varsigma_2}^\dagger d_{\mathbf{k}'+\mathbf{q},\varsigma_2}c_{\mathbf{k}-\mathbf{q},\sigma_1}\right]\,, \tag{17}$$

where $V_{\mathbf{q}}$ is the Coulomb interaction (in a medium with dielectric constant ε_0). Both electrons and holes have parabolic dispersions referred to the bottom of the conduction band and to the top of the valence band respectively.

σ stands for the electron spin $\pm 1/2(\uparrow, \downarrow)$ and ς for the hole spin. We will consider spin effects, later on, only in the 2D case in which only heavy holes with spin $\pm 3/2(\Uparrow, \Downarrow)$ are relevant. For the moment we omit the spin degree of freedom. Then, the number of particles is determined, in thermodynamic equilibrium, by their, spin independent, chemical potentials μ_e and μ_h respectively. The exciton chemical potential is $\mu_X = \mu_e + \mu_h$. The summation of the dispersions of electrons and holes, reproduces the excitons dispersion relation $\epsilon_X(\mathbf{k})$ given in the previous subsection.

In order to describe the condensation in the lowest energy state, created by the operator Ψ_0 given by (7), all the operators as well as the ground state are displaced by

$$D(\sqrt{n_X \mathcal{V}} e^{i\phi}) = e^{\sqrt{n_X \mathcal{V}} e^{i\phi} (\Psi_0^\dagger - \Psi_0)} = \prod_{\mathbf{k}} e^{\sqrt{n_X \mathcal{V}} e^{i\phi} \varphi(\mathbf{k}) (c_{\mathbf{k}}^\dagger d_{-\mathbf{k}}^\dagger - d_{-\mathbf{k}} c_{\mathbf{k}})} . \quad (18)$$

This gives a ground state which is a BCS-like coherent state

$$| \psi_{cohe} \rangle = D(\sqrt{n_X \mathcal{V}} e^{i\phi}) | 0 \rangle = \prod_{\mathbf{k}} \left(u_{\mathbf{k}} - e^{i\phi} v_{\mathbf{k}} c_{\mathbf{k}}^\dagger d_{-\mathbf{k}}^\dagger \right) | 0 \rangle , \quad (19)$$

where

$$u_{\mathbf{k}} = cos[\sqrt{n_X} \varphi(\mathbf{k})] \quad ; \quad v_{\mathbf{k}} = sin[\sqrt{n_X} \varphi(\mathbf{k})] \quad (20)$$

so that $u_{\mathbf{k}}^2 + v_{\mathbf{k}}^2 = 1$. Although $\varphi(\mathbf{k})$ should be determined self-consistently, a good approximation is just to take it as the Fourier-transform of the ground state wave function of a single exciton. In the dilute limit $n_X(a_B)^D << 1$, $u_{\mathbf{k}} \simeq 1$ and $v_{\mathbf{k}} \simeq \sqrt{n_X} \varphi(\mathbf{k})$ which is the exciton wave-function renormalized by the square root of the density. It must be stressed that $u_{\mathbf{k}}$ and $v_{\mathbf{k}}$ are real because both the displacement $\sqrt{n_X \mathcal{V}}$ and the wave function $\varphi(\mathbf{k})$ are also real. We will relax later this condition and allow the existence of a relative phase between the two terms in (19).

The displacement of the fermion operators is the Bogoliubov transformation

$$\begin{aligned} \alpha_{\mathbf{k}} &= u_{\mathbf{k}} c_{\mathbf{k}} + v_{\mathbf{k}} e^{i\phi} d_{-\mathbf{k}}^\dagger \\ \beta_{\mathbf{k}} &= u_{\mathbf{k}} d_{\mathbf{k}} - v_{\mathbf{k}} e^{i\phi} c_{-\mathbf{k}}^\dagger \end{aligned} \quad (21)$$

which are operators referred to the quasiparticles of the system. The Hamiltonian, written in terms of these quasiparticles has three terms:

$$H_X = E_{GS} + H_0 + H_{int} \quad (22)$$

where the first term does not contain quasi-particle operators so that it can be interpreted as the ground state energy

$$E_{GS} = \sum_{\mathbf{k}} [\epsilon_X(\mathbf{k}) - \mu_X] v_{\mathbf{k}}^2 - \sum_{\mathbf{k}, \mathbf{k}'} [v_{\mathbf{k}}^2 v_{\mathbf{k}'}^2 + u_{\mathbf{k}} u_{\mathbf{k}'} v_{\mathbf{k}} v_{\mathbf{k}'}] . \quad (23)$$

The second term contains quadratic dispersion of the quasi-particles

$$H_0 = \sum_{\mathbf{k}} (\alpha_{\mathbf{k}}^{\dagger}\alpha_{\mathbf{k}} + \beta_{\mathbf{k}}^{\dagger}\beta_{\mathbf{k}})$$

$$\times \left\{ \left[\frac{\epsilon_X(\mathbf{k}) - \mu_X}{2} - \sum_{\mathbf{k}'} V_{\mathbf{k}-\mathbf{k}'} v_{\mathbf{k}'}^2 \right] (u_{\mathbf{k}}^2 - v_{\mathbf{k}}^2) + 2u_{\mathbf{k}} v_{\mathbf{k}} \sum_{\mathbf{k}'} V_{\mathbf{k}-\mathbf{k}'} u_{\mathbf{k}'} v_{\mathbf{k}'} \right\}$$

$$- \sum_{\mathbf{k}} (\alpha_{\mathbf{k}}^{\dagger}\beta_{-\mathbf{k}}^{\dagger} + \beta_{-\mathbf{k}}\alpha_{\mathbf{k}})e^{i\phi}$$

$$\times \left\{ \left[\epsilon_X(\mathbf{k}) - \mu_X - 2\sum_{\mathbf{k}'} V_{\mathbf{k}-\mathbf{k}'} v_{\mathbf{k}'}^2 \right] u_{\mathbf{k}} v_{\mathbf{k}} - (u_{\mathbf{k}}^2 - v_{\mathbf{k}}^2) \sum_{\mathbf{k}'} V_{\mathbf{k}-\mathbf{k}'} u_{\mathbf{k}'} v_{\mathbf{k}'} \right\}, \tag{24}$$

while the third term, H_{int}, contains more than two operators and it describes the interactions between the quasi-particles. In a first approximation, this third term can be neglected. In the dispersion of the quasi-particles, H_0, the second term does not conserves the number of particles. As we did for bosons, we make a mean field approach (Hartree-Fock) in which this contribution is canceled out by making zero its coefficient:

$$\left[\epsilon_X(\mathbf{k}) - \mu_X - 2\sum_{\mathbf{k}'} V_{\mathbf{k}} \right] u_{\mathbf{k}} v_{\mathbf{k}} = (u_{\mathbf{k}}^2 - v_{\mathbf{k}}^2) \sum_{\mathbf{k}'} V_{\mathbf{k}-\mathbf{k}'} u_{\mathbf{k}'} v_{\mathbf{k}'} . \tag{25}$$

which is a non-linear equation determining the exciton chemical potential μ_X (another possibility is to fix the exciton chemical potential and use this non-linear condition to determine $\varphi(\mathbf{k})$). Up to first order in $n_X(a_B)^D$, this gives $\mu_X = \epsilon_X(0)(1 - 26\pi n_X(a_B)^D/3)$.

The condition of canceling out the last term in H_0 gives a dispersion relation for the quasi-particles obtained from the first term in H_0:

$$E(\mathbf{k}) = \left[\frac{\epsilon_X(\mathbf{k}) - \mu_X}{2} - \sum_{\mathbf{k}'} V_{\mathbf{k}-\mathbf{k}'} v_{\mathbf{k}'}^2 \right] (u_{\mathbf{k}}^2 - v_{\mathbf{k}}^2) + 2u_{\mathbf{k}} v_{\mathbf{k}} \sum_{\mathbf{k}'} V_{\mathbf{k}-\mathbf{k}'} u_{\mathbf{k}'} v_{\mathbf{k}'}$$

$$= \frac{1}{2u_{\mathbf{k}} v_{\mathbf{k}}} \sum_{\mathbf{k}'} V_{\mathbf{k}-\mathbf{k}'} u_{\mathbf{k}'} v_{\mathbf{k}'} . \tag{26}$$

In terms of the order parameter

$$\Delta(\mathbf{k}) = \sum_{\mathbf{k}'} V_{\mathbf{k}-\mathbf{k}'} u_{\mathbf{k}'} v_{\mathbf{k}'} , \tag{27}$$

the dispersion of the quasi-particles takes the form:

$$H_0 = \sum_{\mathbf{k}} \frac{E(\mathbf{k})}{2} (\alpha_{\mathbf{k}}^{\dagger} \alpha_{\mathbf{k}} + \beta_{\mathbf{k}}^{\dagger} \beta_{\mathbf{k}})$$

$$E(\mathbf{k}) = \sqrt{[\epsilon_X(\mathbf{k}) - 2 \sum_{\mathbf{k}'} V_{\mathbf{k}-\mathbf{k}'} v_{\mathbf{k}'}^2 - \mu_X]^2 + 4\Delta^2(\mathbf{k})}$$

$$u_{\mathbf{k}} = \sqrt{\frac{1}{2} + \frac{\epsilon_X(\mathbf{k}) - 2 \sum_{\mathbf{k}'} V_{\mathbf{k}-\mathbf{k}'} v_{\mathbf{k}'}^2 - \mu_X}{2E(\mathbf{k})}}$$

$$v_{\mathbf{k}} = \sqrt{\frac{1}{2} - \frac{\epsilon_X(\mathbf{k}) - 2 \sum_{\mathbf{k}'} V_{\mathbf{k}-\mathbf{k}'} v_{\mathbf{k}'}^2 - \mu_X}{2E(\mathbf{k})}} . \tag{28}$$

$E(\mathbf{k})$ is the energy required to excite an electron-hole pair with momentum $\mathbf{k}$ out of the condensate. It is a positive quantity which means that a gap exist for the excitation of quasi-particles so that this state is labeled as an "excitonic insulator". The gap is due to the existence of $\Delta(\mathbf{k})$ which is an order parameter as made clear by making the inverse Bogoliubov transformation:

$$\langle c_{\mathbf{k}}^{\dagger} d_{-\mathbf{k}}^{\dagger} \rangle = \langle d_{-\mathbf{k}} c_{\mathbf{k}} \rangle = -u_{\mathbf{k}} v_{\mathbf{k}} e^{i\phi} \neq 0 \tag{29}$$

while the number of condensed excitons is

$$N_X = \sum_{\mathbf{k}} \langle c_{\mathbf{k}}^{\dagger} c_{\mathbf{k}} \rangle = \sum_{\mathbf{k}} \langle d_{\mathbf{k}}^{\dagger} d_{\mathbf{k}} \rangle = \sum_{\mathbf{k}} v_{\mathbf{k}}^2 . \tag{30}$$

Both in the high and low density limits, there are analytic expressions for the dependence of the gap on the density [8, 9]. For high density, the gap decreases exponentially with increasing density while in the low density limit, $\Delta \propto \sqrt{N_X} + O[N_X]$. In this way, one recovers the low density bosonic limit, discussed in the previous subsection, where the excitation spectrum was gapless.

This scheme can be generalized to all the densities by treating the coefficients u and v in terms of a variational description instead of making a perturbation expansion. This gives a good interpolation between the low (excitons) and high (electron-hole pairs) density limits. Figure 1 shows $v_{\mathbf{k}}^2$ for different densities as calculated variationally for a two-dimensional system. For low densities, $v_{\mathbf{k}}$ is simply the Fourier-transform of the exciton wavefunction $\varphi(\mathbf{k})$. When density increases and $n_X(a_B)^2$ approaches 1 (curves e and f in the figure), the fermion character of electrons and holes produces the saturation $u_{\mathbf{k}}^2 + v_{\mathbf{k}}^2 = 1$ and the pair occupation looks like a Fermi distribution function. In the high density limit, the product $u_{\mathbf{k}} v_{\mathbf{k}}$ is non-zero only in a small range of $\mathbf{k}$ around the "Fermi wave vector". These are the only states contributing to the order parameter Δ. Figure 2 shows a scheme of the energy spectrum of an excitonic insulator [11]. This figure clearly suggests an interesting alternative: the binding energy per electron-hole pair can be larger than the initial single particle band gap energy of the semiconductor. In such a case, a permanent excitonic insulator with a finite density of infinite

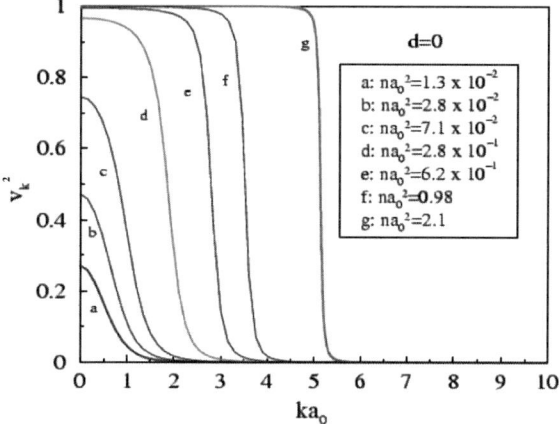

Fig. 1. Ground state fermion distribution $\mid v_k \mid^2$ as a function of electron-hole pair internal kinetic energy for various densities in a 2D system

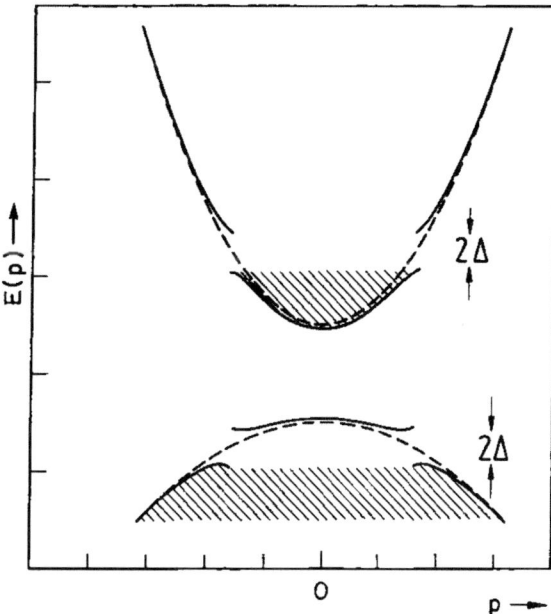

Fig. 2. Sketch of the energy spectrum of the condensed exciton phase. Hatched regions depict the occupation by electrons

life-time excitons is the actual ground state of the system. This implies the spontaneous reconstruction of the electronic structure of the crystal.

Hitherto, we have treated the case of zero temperature and obtained a finite order parameter Δ. Since we have used a BCS-like description, the behavior of the system with temperature must be similar to the one of superconductors (second order phase transition which becomes first order in the presence of a magnetic field). Increasing temperature reduces the excitation gap which becomes zero at the critical temperature for condensation $T_c \propto \Delta(T = 0)$. As in the case of BEC or KT transitions, typical parameters for excitons imply a critical temperature of a few Kelvins.

The mean field description here described predicts, among other properties, an increase of frequency of emitted light with respect to the non-condensed case. This prediction does not agree with experimental evidence for very low temperatures. This is mainly due to the use of a simple mean field approach which neglects correlation effects. The simplest way of taking these effects into account is to consider a screened interaction (e.g. in RPA). We do not describe here this approach developed by Nozieres and coworkers.

2.4 Electron-Hole Liquid

When the density of electrons and holes is very high, screening can be so efficient that electron-hole attraction is not intense enough for producing pairing. A dense plasma with two components, electrons and holes, is the most probable ground state of the system. It is possible to show that this is the actual ground state for any density of electron-hole pairs when the interaction between these pairs is attractive. However, such interaction is in general repulsive so that a threshold density is required for this Mott transition between the condensed pairs state and the electron and hole liquid.

3 Emission of Light

Since a gap in the excitation spectrum is expected, a good way of experimentally determining the adequacy of the theory would be to measure the gap. Unfortunately, no experimental evidence of such a gap exists.

Using the large experience of people working in quantum optics, the best way of detecting condensation, is to determine the existence of coherence by measuring correlation functions. In particular, second order coherence function as done in intensity correlation experiments of the Hanbury Brown–Twiss type. Since one can not directly measure properties of the excitonic field $\Psi(\mathbf{r}, t)$, one must perform experiments with the photon field emitted when excitons recombine. An additional difficulty with these kind of experiments is that, in many cases, typical time scales are smaller than experimental resolution of apparatus. Since only in the case of polaritons, to be discussed later, these kind of Hanbury Brown–Twiss experiments have been carried out, we postpone the discussion of such technique and analyze the question of light emission.

The coupling of excitons to a photonic field implies that an exciton condensate can emit coherent light [12, 13]. The importance of this is twofold:

the coherent emission could be used to probe exciton condensation and, the emission from the condensate can occur in the absence of population inversion [12]. Now the total Hamiltonian of the photons plus the excitons is $H = H_X + H_L + H_{XL}$ where H_X is the excitonic part previously discussed, and

$$H_L = \sum_{\mathbf{q}} cq b_{\mathbf{q}}^{\dagger} b_{\mathbf{q}}, \tag{31}$$

$$H_{XL} = gD_{cv} \sum_{\mathbf{k},\mathbf{q}} c_{\mathbf{k}}^{\dagger} d_{-\mathbf{k}}^{\dagger} b_{\mathbf{q}} + h.c. \tag{32}$$

We have not specified the angular momenta of electrons and holes as well as about the polarization of the light but the obvious selection rule is implicit. $b_{\mathbf{q}}^{\dagger}$ are the photon operators with momentum $\mathbf{q}$ along the direction perpendicular to the 2D system due to conservation of in-plane momentum in (32). H_{XL} is the exciton-light dipolar coupling given in the rotating wave approximation. H_{XL} describes processes in which one exciton is created and one photon is destroyed or vice versa; the total number of excitons *plus* photons is conserved. g is the light-matter coupling constant and D_{cv} is the matrix element of the dipole moment operator between the valence and the conduction band.

There is an electric dipole moment associated with the exciton gas:

$$P \equiv \sum_{\mathbf{k}} D_{cv} c_{\mathbf{k}}^{\dagger} d_{-\mathbf{k}}^{\dagger} + h.c. \tag{33}$$

In the absence of condensation, $\langle P \rangle$ is equal to zero. On the contrary, in the case of the condensate, the finite order parameter $\langle c_{\mathbf{k}}^{\dagger} d_{-\mathbf{k}}^{\dagger} \rangle$ produces a non zero average dipole

$$\langle P \rangle = -2D_{cv} cos(\phi) \sum_{\mathbf{k}} u_{\mathbf{k}} v_{\mathbf{k}} \tag{34}$$

which oscillates with time [14]. A simple way of getting this oscillation is given by [15]

$$\frac{d}{dt}\phi = \frac{\partial}{\partial N_X} E(N_X, \phi) \tag{35}$$

where $E(N_X, \phi) = \langle H_X + H_{XL} \rangle$. By analyzing the main contributions to this energy [13], one gets a time dependent electric dipole moment

$$\langle P(t) \rangle = -2D_{cv} cos\left(\epsilon_X(0)t + \phi(0)\right) \sum_{\mathbf{k}} u_{\mathbf{k}} v_{\mathbf{k}} \tag{36}$$

which produces an oscillating electromagnetic field.

The quantum description of the light emitted by the condensate is made with the part of the Hamiltonian which contains photon operators, in which the polarization operator is substituted by its time dependent statistical average

$$H_L + H_{XL} = \sum_{\mathbf{q}} \left[cq b_{\mathbf{q}}^{\dagger} b_{\mathbf{q}} - \frac{g D_{cv} \sum_{\mathbf{k}} u_{\mathbf{k}} v_{\mathbf{k}}}{2} \left(e^{-i\phi(t)} b_{\mathbf{q}} + h.c. \right) \right]. \quad (37)$$

This Hamiltonian describes a set of non interacting forced harmonic oscillators in a Glauber coherent state [13]

$$|\Xi(t)\rangle = e^{-iH_L t} \prod_{\mathbf{q}} e^{\Theta_q(t)} e^{iK_q(t) b_{\mathbf{q}}^{\dagger}} |0\rangle \quad (38)$$

where

$$K_q(t) = \frac{g}{2} \int_0^t D_{cv} \sum_{\mathbf{k}} u_{\mathbf{k}} v_{\mathbf{k}} e^{i[cqs-\phi(s)]} ds; \quad \Theta_q(t) = \int_0^t K_q(s) \frac{dK_q^*(s)}{ds} ds. \quad (39)$$

Equation (38) proves the coherence of the emission coming from an exciton condensate.

The coupling of excitons with photons has another very interesting consequence: The emission of light stabilizes the condensation process allowing the existence of an exciton condensate at finite temperatures even in a perfectly flat infinite 2D system. The prove that condensation is impossible in 2D systems at finite temperatures relies on the fact that thermal fluctuations of the phase of the condensate provoke, in the momentum distribution, an infrared singularity of the form [16, 17]

$$\int \frac{d^2 \mathbf{k}}{\langle [[\rho_{\mathbf{k}}, H_X], \rho_{\mathbf{k}}^{\dagger}] \rangle} = \int \frac{m_X d^2 \mathbf{k}}{k^2 N_0} \quad (40)$$

where $\rho_{\mathbf{k}}$ is the $\mathbf{k}$ Fourier component of the density operator. However, light emission, adds to the denominator the term

$$\langle [[\rho_{\mathbf{k}}, H_{XL}], \rho_{\mathbf{k}}^{\dagger}] \rangle = g D_{cv} \sum_{\mathbf{q},\mathbf{p}} \langle b_{\mathbf{q}}^{\dagger} \rangle \left(\langle c_{\mathbf{p}+\mathbf{k}}^{\dagger} d_{-\mathbf{p}+\mathbf{k}}^{\dagger} \rangle \right.$$
$$\left. + \langle c_{\mathbf{p}-\mathbf{k}}^{\dagger} d_{-\mathbf{p}-\mathbf{k}}^{\dagger} \rangle - 2 \langle c_{\mathbf{p}}^{\dagger} d_{-\mathbf{p}}^{\dagger} \rangle \right) + h.c. \quad (41)$$

which depends linearly on k and suppresses the singularity. Long wavelength thermal fluctuations do not destroy the long range order because the condensed excitons can recombine emitting coherent light. So, condensation of optically active excitons is possible in 2D at finite temperatures. With respect to the critical temperature, one can give upper bounds. There are two relevant energy scales which are upper bounds for T_c[3]: the BCS gap Δ and the quantum degeneracy temperature, $T_{QD} = 2\pi\hbar^2 n/m_X$. Within the BCS approximation [18], $\Delta > T_{QD}$ for densities below 10^{11} cm^{-2} and interlayer spacing between electrons and holes smaller than 100 Å. For typical GaAs quantum wells a typical $T_c = T_{QD}$ results to be a few degrees K.

In general, the available experimental information refers to the emitted spectrum which is the Fourier transform of the first order correlation function

$$S(\omega) = \frac{1}{\pi} \Re \int_0^\infty d\tau e^{i\omega\tau} \langle b_0^\dagger(t+\tau)b_0(t)\rangle \qquad (42)$$

where we have considered that light is emitted perpendicular to the 2D system, i.e. $\mathbf{q} = 0$. The spectrum is usually measured by continuous wave spectroscopy in the stationary limit $(t \to \infty)$ or, sometimes, by means of ultrafast spectroscopy (as a function of t). Pioneering experiments [2, 1] were performed in Cu_2O, a 3D system. However, the main experimental effort has been devoted to quasi-2D systems due to the great quality achieved in its fabrication. Spectral narrowing, enhanced scattering and long range transport in non-luminescence states have been reported [1, 20, 21]. However, none of these experiments has been able to clearly establish the existence of condensation.

Another relevant quantity for complementing the measurement of the spectrum of the light emitted by the electron-hole condensate is the second-order coherence, $g^{(2)}(\tau)$, that we will discuss more carefully in the case of polaritons. $g^{(2)}(\tau)$ presents interesting signatures of the transition between a BCS state and a BEC of excitons in a range of delays (τ) too short for the current experimental capabilities [19].

4 Magnetoexcitons

Since many years ago [22] it was realized that the presence of a magnetic field introduces interesting effects in the properties of excitons. The kinetic energy perpendicular to the field, becomes quantized in Landau levels. So, 2D systems in a perpendicular magnetic field have a discrete spectrum with the same finite degeneracy for each state. Moreover, in a Landau gauge, the one-dimensional k-vector is directly associated with a center of orbit. Therefore, the electrons and holes with a given k are spatially localized in the same region increasing the Coulomb attraction responsible for the formation of the magnetoexciton. This improves the possibility of spontaneous coherences in mean field states describing combinations of magnetoexcitons. Apart from pairs in which states from valence and conduction bands are involved, a generalization of intraband magnetoexcitons can be built up by mixing conduction band states in different Landau levels, in the same Landau level with two different spins, in different layers of a bilayer system or any other combination involving two values of a given degree of freedom. BCS-like wave functions involving different kinds of magnetoexcitons have been proposed to understand different experimental phenomena in the quantum Hall regime as skyrmions close to filling factor 1, merons in bilayers, Coulomb drag or magnetotunneling in bilayers, etc.[23]. Due to the coherence involved by these mean field wave functions, one can interpret all these processes as evidence of the condensation of those generalized magnetoexcitons.

5 Multicomponent Condensates

Hitherto, we have ignored the fact that excitons have an internal degree of freedom: the third component M of the angular momentum which is, improperly, labeled as the spin of the exciton. This opens an extremely interesting new possibility: the existence of multicomponents condensates as those occurring in some superfluid ^{3}He phases [24]. Since only excitons with $M = \pm 1$ are optically active, these are the only ones of interest for us (Dark excitons can play an important role in the dynamics of the whole exciton system and be essential in the understanding of some experiments). The excitation or recombination of these two types of excitons occurs through (right/left) circularly polarized light allowing a good control and characterization of the different components of the system. It appears a spin polarization $P_S = (N_+ - N_-)/(N_+ + N_-)$ where $N_\pm$ are the populations of excitons with $M = \pm 1$ respectively. The system can be ferromagnetic when $N = N_+$, paramagnetic when $N_+ = N_-$ or some intermediate state and it can support Josephson-like oscillations between the two components of the condensate.

5.1 Phase Diagram

Since one of the problems in exciton condensation is the lifetime of the excitons, one can make it longer by separating electrons from holes. With this motivation, we are going to consider a bilayer system with electrons in one layer and holes in the other with an interlayer distance d. It turns out that the most interesting spin effect, that is spin polarization, of the exciton gas depends on d. The Hamiltonian is the complete Hamiltonian (17) but now all the wave vectors are 2D. We keep all the dependences on the spin ($(\uparrow, \downarrow)$ for electrons and $(\Uparrow, \Downarrow)$ for holes) and, on top of this, the interaction between the different fermions depends explicitly on the interlayer distance d: $V(r, d) = -e^2/\varepsilon_0 \sqrt{r^2 + d^2}$ with r being the modulus of an inplane vector.

Following the mean field approach described above for the case without spin, one can write a BCS-like variational wave function as done for the case of A phase of the superfluid ^{3}He [24]

$$|\Psi(\phi_+, \phi_-)\rangle = \prod_{\mathbf{k},\mathbf{k}'} \left(u_k^+ + e^{i\phi_+} v_k^+ c_{\mathbf{k}\downarrow}^\dagger d_{\mathbf{k}\Uparrow}^\dagger \right) \left(u_{k'}^- + e^{i\phi_-} v_{k'}^- c_{\mathbf{k}'\uparrow}^\dagger d_{\mathbf{k}'\Downarrow}^\dagger \right) | 0 \rangle . \quad (43)$$

Due to the rotational symmetry in the 2D plane, the parameters u and v only depend on the modulus of vector $\mathbf{k}$. This wave function is the product of two BCS states each one describing a condensate of excitons with $M = \pm 1$. The variational parameters $u_k^\pm$ and $v_k^\pm$ can be determined numerically [25] looking for an extremal of the Hamiltonian with constrains on the number of particles $N_\pm$ and the chemical potentials $\mu_\pm$ for the two components of the condensate. For given values of $\mu_\pm$, such extremal, does not depend on $\phi_\pm$. From $v_k^\pm$ one gets the density, $n_\pm = \sum_k (v_k^\pm)^2$, for each type of exciton. The variational procedure provides $\mu_\pm$ as a function of $n = n_+ + n_-$ and $n_r = n_+ - n_-$ for a certain value of d. The result takes the form $\mu_\pm = \epsilon_X(0) + \bar\mu_\pm(n, n_r)$. $\bar\mu_\pm(n, n_r)$ contain three many exciton contributions [25]: the first is a Hartree term giving the coupling between the two components of the condensate. The other two contributions are related to the Pauli exclusion principle. One is the exchange correction (EC), i.e. the reduction of the repulsive electron-electron (hole-hole) interaction due to the fact that two identical fermions can not occupy the same state. The last contribution is the vertex correction (VC), which represents the reduction of the e-h attraction due to the occupation of final states in e-h scattering processes. EC and VC contributions to the chemical potential of a fermion with a given spin are independent of the amount of fermions with the opposite spin. This predicts a non zero splitting, $\delta = \mu_+ - \mu_-$, whenever there is a non zero n_r.

The predictions of this approach could be tested by a photoluminescence experiment. A bilayer system can be excited with circularly polarized light which creates a density of, for instance, n_+ excitons which decreases by both recombination and by spin relaxation, resulting in a population of $M = -1$ excitons with density n_-. Although both n and n_r evolve in time, exciton can be considered in a quasi-equilibrium state. $\mu_\pm$ are measurable by the frequency of the photons emitted with the two possible circular polarizations. The intensities of these emission lines give information about n and n_r so that theoretical predictions can be tested experimentally.

The spin dependence of EC and VC produces splitting, $\delta \equiv \mu_+ - \mu_-$, whenever there is a non zero spin polarization P as shown in the inset of Fig. 3. δ is an increasing function of P and, for moderate values n, is also an increasing function of n. δ is in the range of meV and is consequently an experimentally accessible quantity.

An extremely interesting physical property connected with the *sign* of δ is that it determines the most stable state of the system. The total energy E_{TOT} has the following property in the thermodynamic limit [25]:

$$\left(\frac{\partial E}{\partial P}\right)_N \equiv E(N_+ + 1, N_- - 1) - E(N_+, N_-) = \delta . \tag{44}$$

Positive δ means $n_+ > n_- \Rightarrow \mu_+ > \mu_-$, i. e. the majority ($M = +1$) excitons are less bounded. Negative δ means that, whenever $n_+ > n_- \Rightarrow \mu_+ < \mu_-$, i. e. the majority ($M = +1$) excitons are more bounded. For given n and d, the energy is a monotonous function of P. $\delta > 0$ corresponds to an increase of the total energy with P, and the minimum energy state is the unpolarized one, i.e. $P_{me} = 0$. $\delta < 0$ indicates that the system should have minimum energy when completely polarized, $P_{me} = 1$.

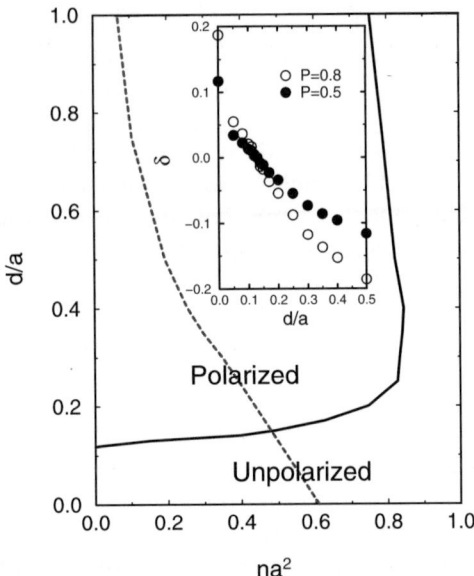

Fig. 3. Phase diagram for both sign of $\delta = \mu_+ - \mu_-$ and polarization n_r/n of the ground state. The continuous line separates the region of $\delta < 0$ (spin polarized) and $\delta > 0$ (spin unpolarized). The region on the *left* (*right*) of the *dashed line* corresponds to a condensate of bounded (unbounded) $e - h$ pairs. The inset shows the splitting $\delta = \mu_+ - \mu_-$, in units of $|E_0^{2D}|$, as a function of the $e - h$ separation d for $na^2 = 0.15$ and $n_r/n = 0.5$ (and 0.8)

The physical origin of the dependence of sign of δ on n and d resides in that VC gives a positive contribution to δ while EC gives a negative contribution. At high n, the kinetic energy dominates any interaction effect and δ is positive for any d. For low n, δ will be positive if EC is larger than VC and negative otherwise. For a fixed value of n, the ratio between VC and EC varies as d is changed. As d increases, VC decreases while EC increases as a consequence of a larger exciton size, this implies an increase of the overlap between excitons. For moderate densities, the change in δ's sign occurs at a critical value $d_{cr} \simeq 0.12a \simeq 10\,\text{Å}$ practically independent on P. These results are presented in the phase diagram of Fig. 3.

5.2 Coherence Effects

We have shown already that an exciton condensate emits coherent light. The existence of two components, coupled to each other, in the condensate opens new interesting possibilities, related to Josephson-like effects, particularly with respect to the light emission.

Exciton condensation is described by the BCS-like wave function (43) containing two different phases phases $\phi_\pm$ each of them being the same for all the $\pm$-like excitons. This is coherence and it has a macroscopic implication: the mean value of the electric dipole operator, $P_\pm$, associated to $\pm$ excitons, is not zero for a condensate.

$$\langle P_\pm \rangle = \langle \sum_{\mathbf{k},\sigma,\varsigma} \overline{D_{cv}(\pm)} c^\dagger_{\mathbf{k},\sigma} d^\dagger_{-\mathbf{k},\varsigma} + h.c. \rangle = 2 \sum_k u^\pm_k v^\pm_k cos(\phi_\pm) \qquad (45)$$

where $|\phi_\pm\rangle$ are those of (43) and $\overline{D_{cv}(\pm)} = D_{cv}\delta_{\pm,\sigma+\varsigma}$ is the electric dipole matrix element for an electron-hole pair. Time evolution of $\langle P_\pm \rangle$ is given by the phase equation of motion [15]

$$\langle P_\pm(t) \rangle = 2\overline{D_{cv}(\pm)} \sum_k u^\pm_k v^\pm_k cos\left(\phi(0) + \mu_\pm t\right) . \qquad (46)$$

As we discussed in previous sections, exciton condensation is associated with an oscillating electric polarization which produces the emission of coherent (circularly polarized) light. Now, the two components of the condensate have slightly different macroscopic polarizations oscillating at two frequencies $\mu_\pm$ each of them close to $\epsilon_X(0)$ but having opposite (left or right) circular polarizations. The polarization of the coherent light emitted by a two component condensate depends on the relative phase of the condensates. In the previous subsection we saw that the picture of two exciton gases coexisting in the same spatial region, and essentially uncoupled trough many body interactions is quite reasonable. In case of condensation, any tiny mechanism, H_c that couples the $M = +1$ to the $M = -1$ excitons will have enormous effects. H_c could be, for instance, the inter excitonic exchange [26, 27] which has a mean value $g(d)N\cos(\phi_+ - \phi_-)$ in the state (43).

Following the ideas involved in the study of the Josephson effect [15] in systems with two coexisting condensates [24, 28], the relevant physical variable are $\phi_r \equiv \phi_+ - \phi_-$ and N_r. These operators are conjugated magnitudes from each other. The effective Hamiltonian governing the behavior of ϕ_r and N_r is obtained from the average of the true microscopic Hamiltonian in the state (43):

$$\langle H_{eff}(\phi_r, N_r) \rangle \equiv \langle H + H_c \rangle = \frac{c(d)a_B^2}{S}(N_r)^2 + g(d)N\cos(\phi_r) + \mathcal{F}(N) \quad (47)$$

where the first term comes from interaction effects discussed in the previous subsection. S is the area of the 2D system. Hamiltonian (47) is like a pendulum of frequency $\Omega = 2\sqrt{c(d)gna_B^2}$, when one interprets N_r as the generalized momentum and ϕ_r as the generalized coordinate. Neglecting quantum fluctuations, ϕ_r and N_r are classical quantities. Finally the last term is simply an additive constant because it does not depend on N_r and ϕ_r. As the first term is divided by an extensive quantity, S, and the $cos(\phi_r)$ is multiplied by another extensive quantity, N, the important term in $\langle H_{eff} \rangle$ is that of the relative phase which will be obtained by minimizing $\langle H_{eff} \rangle$. $g < 0$ gives

$\phi_r = 0$ while $g > 0$ gives $\phi_r = \pi$. This can be experimentally checked. The nonzero components of the macroscopic electric polarization are those parallel to the bilayer. When the condensate has two components with the same density, i.e. $n_r = 0$, one gets $\mu_\pm \equiv \mu$, $v_k^\pm = v_k$ and $u_k^\pm = u_k$ and one obtains Cartesian components:

$$P_x \simeq \left(\sum_k u_k v_k\right) \cos(\phi_r/2) \cos(\mu t + \phi(0))$$
$$P_y \simeq \left(\sum_k u_k v_k\right) \sin(\phi_r/2) \cos(\mu t + \phi(0)) \,. \tag{48}$$

As a consequence of the relative phase, instead of circularly polarized emission, the emitted light will be linearly polarized in the x or y axis, depending on the sign of $g(d)$. A tiny coupling between the $M = +1$ and the $M = -1$ components of the condensate gives rise to the locking of the relative phase. In a general case, $n_r \neq 0$ implies the light to be elliptically polarized and with a beating of frequency δ in the intensity of the emitted signal due to the classical interference of the $M = +1$ and $M = -1$ emitted fields. The experimental observation of the linearly or elliptically polarized light would be much simpler than that of the coherent nature of the emitted light.

6 Polariton Condensation

As mentioned before, in spite of many efforts to experimentally detect exciton condensation no clear evidence of such transition has been found up to the present. But all the different proposals made for such task give insight for a new possibility: since the light emission seems to be the best experimental alternative, instead of thinking in a condensate of excitons that, subsequently emits coherent photons, one can think in a system in which excitons and photons are strongly coupled giving a different quasi-particle, labeled as polariton, that can condensate itself. In this way, one could measure directly the properties of the condensate because the photonic component of the polariton could be easily accessible to experimental analysis. These kind of quasi-particles can be actually obtained by confining a 2D material system supporting excitons (a QW) within an optical microcavity (mirrors in just one spacial direction) with normal optical modes close to resonance with the excitons. Apart from other practical questions, this alternative of polariton condensation has a small problem and a big advantage. The problem is that cavity photons can escape from the system reducing the life-time of the polaritons they form in one order of magnitude compared with life-time of excitons. Still, such polariton life-time is large enough (a few ps) to allow thermalization and condensation of polaritons. On the other side, a big advantage of polaritons resides in its low effective mass (typically 10^{-4} times that of an exciton) which implies a higher critical temperature and, more important, lower population threshold for condensation. This question of the threshold density will attract a large part of our interest.

Bose–Einstein condensation of polaritons was first proposed by A. Imamoglu et al. [12], and there has been a lot of theoretical [29] and experimental [30] activity on this topic since then. A clear manifestation of the current interest of polariton condensation is the fact that, at difference with excitons, this system has been experimentally studied [30] by means of a Hanbury Brown–Twiss experiment with the intention of measuring second order coherent function:

$$g^{(2)}(t, t + \tau) = \frac{\langle E^{(-)}(t) E^{(-)}(t + \tau) E^{(+)}(t + \tau) E^{(+)}(t) \rangle}{\langle E^{(-)}(t) E^{(+)}(t) \rangle^2}. \tag{49}$$

where $E^{(-)}$ and $E^{(+)}$ are the negative and positive frequency parts of the electric field operator, respectively. A certain average of this magnitude has been measured in a microcavity excited by a pulsed laser. Apart from many practical difficulties related with large time (ns) averages due to the use of avalanche detectors, the experimental output is $g^{(2)}(\tau = 0)$, a magnitude which takes very different values for different types of quantum states. For a chaotic state involving Gaussian distributions of bosons, $g^{(2)}(0) = 2$, while for a coherent state involving Poisson distributions of bosons, $g^{(2)}(0) = 1$ ($g^{(2)}(0) < 1$ implies subpoissonian distributions of bosons). Although a statistical mixture of photon number states with Poissonian distribution is not coherent but it also has $g^{(2)}(0) = 1$, any experimental measure of the $g^{(2)}(0)$ giving a value significantly lower than 2 is usually considered as a good indication of coherence of the field (a laser just above threshold has $g^{(2)}(0) = \pi/2$). In the mentioned experiment in GaAs microcavities, $g^{(2)}(0)$ is close to 2 for low pumping densities while it decreases up to a value around 1.5, not being possible to increase the density of particles because filling of fermionic phase space would start playing an important role.

6.1 Polariton Dynamics

Excitons are always coupled to photons, but usually this coupling is weak enough to be considered as a perturbation that leads, for example, to the radiative decay of the electron-hole pair. If the exciton-photon coupling is large enough (strong coupling regime), then the whole exciton spectrum is renormalized and one can no longer speak about excitons and photons as separated entities. Instead, polaritons appear as the quasi-particles of the system. This condition is met in the case of a quantum well embedded in a semiconductor microcavity. The microcavity is created by two semiconductor mirrors (Bragg reflectors), and the quantum well is placed between these two mirrors. Photons are confined in the growth direction, and are free to move in the plane of the quantum well only. One can say that both photons and excitons live in 2D. The Hamiltonian that describes excitons, photons, and their coupling, is:

$$H_0 = \sum_{\mathbf{k}} \left[\epsilon_X(\mathbf{k})\Psi_{\mathbf{k}}^{\dagger}\Psi_{\mathbf{k}} + \epsilon_c(\mathbf{k})b_{\mathbf{k}}^{\dagger}b_{\mathbf{k}} + \frac{\Omega_P}{2}\left(b_{\mathbf{k}}^{\dagger}\Psi_{\mathbf{k}} + \Psi_{\mathbf{k}}^{\dagger}b_{\mathbf{k}} \right) \right] \qquad (50)$$

$\epsilon_c(\mathbf{k})$ being the dispersion relation of cavity photons and $\mathbf{k}$ the linear momentum in the plane of the quantum well. Ω_P is the coupling energy (polariton splitting). Since each exciton interacts only with the photon that has the same in-plane momentum, one can diagonalize the Hamiltonian (50) by the linear transformation:

$$a_{LP,\mathbf{k}} = X_{\mathbf{k}}^{LP}\Psi_{\mathbf{k}} + C_{\mathbf{k}}^{LP}b_{\mathbf{k}}$$
$$a_{UP,\mathbf{k}} = X_{\mathbf{k}}^{UP}\Psi_{\mathbf{k}} + C_{\mathbf{k}}^{UP}b_{\mathbf{k}} \qquad (51)$$

where X^j and C^j are called Hopfield coefficients. Original excitons and photons transform now into lower $a_{LP,\mathbf{k}}$, and upper $a_{UP,\mathbf{k}}$ polaritons:

$$H_0 = \sum_{\mathbf{k}} \left(\epsilon_{\mathbf{k}}^{LP}a_{LP,\mathbf{k}}^{\dagger}a_{LP,\mathbf{k}} + \epsilon_{\mathbf{k}}^{UP}a_{UP,\mathbf{k}}^{\dagger}a_{UP,\mathbf{k}} \right). \qquad (52)$$

The lower polaritons are the states in which Bose-Einstein condensation can take place. One can neglect, thus, the upper polariton branch, and from now on the polariton operators $a_{\mathbf{k}}$ will refer always to lower polaritons. It turns out that the photons in a microcavity have a very steep dispersion relation, and polaritons share this property with the original photons. In Fig. 4 dashed lines show the dispersion relations for uncoupled excitons and photons, while continuous line stands for the lower polaritons. Near $\mathbf{k} = 0$ the exciton-photon coupling is very strong and the polaritons show a pronounced dip in momentum-space. This dip is the region of low energy and small mass, where the conditions for polariton condensation are so favorable.

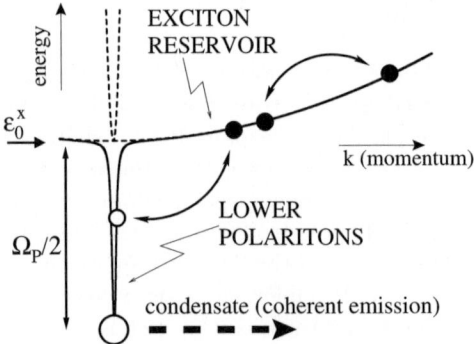

Fig. 4. Phase space for the scattering of quasi-particles in a semiconductor microcavity. Upper and lower *dashed lines* represent the bare photon and bare exciton, respectively, in the case of zero detuning. The continuous line is the polariton dispersion that results from the strong exciton-photon coupling

Polaritons are interacting quasiparticles. The main contribution to this interaction comes from the exciton-exciton interaction due to the Coulomb interaction between electrons and holes forming excitons. The saturation of the exciton-photon oscillator strength is an important physical effect that can be expressed in terms of an effective exciton-photon interaction. We incorporate all these effects in an interaction Hamiltonian:

$$H_I = \sum_{\mathbf{k_1},\mathbf{k_2q}} \frac{M_{XX}}{2S} \Psi^\dagger_{\mathbf{k_1}} \Psi^\dagger_{\mathbf{k_2}} \Psi_{\mathbf{k_1+q}} \Psi_{\mathbf{k_2-q}} + \frac{\sigma_{sat}}{S} \Psi^\dagger_{\mathbf{k_1}} \Psi^\dagger_{\mathbf{k_2}} \Psi_{\mathbf{k_1+q}} b_{\mathbf{k_2-q}} + h.c. \quad (53)$$

where S represents the quantization area, $M_{XX} \approx 6\epsilon_X(0)a_B^2$ and $\sigma_{sat} \approx 1.8\Omega_P a_B^2$. H_I, expressed in the polariton basis, involves an effective polariton-polariton interaction $U_{\mathbf{k_1},\mathbf{k_2},\mathbf{k_3k_4}}$ that depend on M_{XX}, σ_{sat}, and the Hopfield coefficients:

$$H_I = \sum_{\mathbf{k_1},\mathbf{k_2},\mathbf{k_3},\mathbf{k_4}} \frac{1}{2S} U_{\mathbf{k_1},\mathbf{k_2},\mathbf{k_3k_4}} \, a^\dagger_{\mathbf{k_1}} a^\dagger_{\mathbf{k_2}} a_{\mathbf{k_3}} a_{\mathbf{k_4}} \quad (54)$$

The study of polariton condensation requires an approach different to that is usually taken in the case of exciton condensation. The small effective mass of polaritons results in a very high temperature for BEC (of the order of 50 K). However the lifetime of the polaritons is very short, (a few ps), due to the probability for the photonic component to escape through the mirrors of the microcavity. So, the first question to be answered is whether polaritons have enough time to thermalize and condense before they disappear.

In the following we will show that, under certain conditions, the answer to this question is positive. In many experiments, a non-resonant laser produces a population of high energy excitons that, subsequently, relax to the region of strong exciton-photon coupling (the dip in Fig. 4). The density of excitons that is non-resonantly created has some limitations. If the density is above a given value, then the exciton oscillator strength is quenched completely due to the screening by other excitons, and the whole polariton picture makes no sense. This maximum density is called saturation density, $n_{sat} \approx 0.15 a_B^2$ [31]. Therefore, it is better to remain restricted to polariton or exciton densities such that $n \ll n_{sat}$. We will discuss the case of CdTe microcavities which are better candidates for condensation than microcavities based on InGaAs or GaAs. The reason is that for CdTe QW's the exciton binding energy is of the order of 25 meV, and a_B is below 50 Å. This implies a high saturation density 6.7×10^{11} cm^{-2} [32], compared with the case of InGaAs and GaAs QW's, where saturation densities are 6.6×10^{10} cm^{-2} and 1.3×10^{11} cm^{-2}, respectively, thus making more difficult to observe polariton condensation in the strong coupling regime.

In just a few fs after the non resonant excitation of the system, excitons relax to the bottleneck region by emitting optical phonons [33]. In order to describe satisfactorily the polariton dynamics and to predict the possibility of BEC, one can make the following assumptions:

(1) The exciton population above the bare exciton energy can be considered as a reservoir at a given temperature T_X, that follows a Maxwell-Boltzmann distribution. This is a good approximation due to the fast exciton-exciton scattering that thermalize the exciton population at large densities [34].

(2) The main scattering mechanism is the exciton-polariton scattering (shown in Fig. 4), in which two excitons scatter in the reservoir, and the final states are a lower polariton, and another reservoir exciton. This relaxation process is provoked by the polariton-polariton interaction, that can be calculated from the exciton-exciton, and exciton-photon interactions in (53). The scattering of excitons with acoustical phonons going into lower polariton states is also much slower than the exciton-polariton scattering already at exciton densities of $10^9 \, \mathrm{cm}^{-2}$.

Following an approach very successful in the description of BEC of atoms [35] the non-condensed quasiparticles (high energy excitons) are described as a reservoir while only a detailed study of the condensed polaritons is performed. The phase-space is, thus, naturally divided into two pieces:

(i) The exciton reservoir. Taking the energy origin at the bare exciton energy, the exciton reservoir includes all the levels with $\epsilon_k^{LP} < 0$). Exciton and photon levels in the LPB are strongly coupled, and radiative losses are much more important, due to the strong photonic weight of these states. A nonequilibrium polariton distribution that depends strongly on temperature and density can be expected. On the other hand, as mentioned before, the abrupt dispersion relation implies a very small effective mass for the LPB states, $m_{LP} \approx 4 \times 10^{-5} m_X$, leading to the possibility of BEC.

The dynamics of the system is given by a set of rate equations for N_k^{LP} (occupation numbers in the lower polariton branch), and n_X (the density of the thermalized exciton reservoir). Here we are interested on the evolution of occupation numbers, that is, the incoherent dynamics, and the semiclassical Boltzmann equation is enough for this purpose. In the following section we present a derivation of this equation by means of the study of the polariton density matrix. The semiclassical Boltzmann equation [32] is simplified by the fact that we are considering that high energy excitons are thermalized:

$$\frac{dN_k^{LP}}{dt} = W_k^{in} n_X^2 (1 + N_k^{LP}) - W_k^{out} n_X N_k^{LP} - \Gamma_k^{LP} N_k^{LP}$$

$$\frac{dn_X}{dt} = -\frac{1}{S} \sum_k \mathcal{D}_k^{LP} (W_k^{in} n_X^2 (1 + N_k^{LP}) - W_k^{out} n_X N_k^{LP})$$

$$- \Gamma_X n_X + p_X \tag{55}$$

where Γ_k^{LP} and Γ_X are the radiative losses in the LPB and exciton reservoir, respectively. Γ_k^{LP} is proportional to the inverse of the lifetime for the bare photon within the cavity τ_{ph}, typically of the order of ps. p_X represents the non-resonant pump that injects excitons directly into the exciton reservoir. k refers now to the modulus of the 2D momentum $\mathbf{k}$. $\mathcal{D}_k^{LP}$ is the degeneracy for the polariton level corresponding to the wave-vector k. $W_k^{in} n_X^2$, $W_k^{out} n_X$ are the rates for exciton-polariton scattering into and out of the LPB [32]:

$$W_k^{in} n_X^2 = \sum_{\mathbf{k}_2, \mathbf{k}_3, \mathbf{k}_4} 4|U_{\mathbf{k}, \mathbf{k}_2, \mathbf{k}_3, \mathbf{k}_4}|^2 (1 + N_{k_2}^X) N_{k_3}^X N_{k_4}^X \delta(\epsilon_k^{LP} + \epsilon_X(k_2)$$
$$-\epsilon_X(k_3) - \epsilon_X(k_4)) \tag{56}$$
$$W_k^{out} n_X = \sum_{\mathbf{k}_2, \mathbf{k}_3, \mathbf{k}_4} 4|U_{\mathbf{k}, \mathbf{k}_2, \mathbf{k}_3, \mathbf{k}_4}|^2 N_{k_2}^X (1 + N_{k_3}^X)(1 + N_{k_4}^X) \delta(\epsilon_k^{LP} + \epsilon_X(k_2)$$
$$-\epsilon_X(k_3) - \epsilon_X(k_4))$$

where N_k^X are the occupation numbers in the exciton reservoir. The rates W_k^{out}, W_k^{in} have a rather curious dependence on the i temperature of the exciton reservoir, T_X:

$$W_k^{in} \propto e^{-|\epsilon_k^{LP}|/k_B T_X}, \qquad W_k^{out} \propto e^{-2|\epsilon_k^{LP}|/k_B T_X}. \tag{57}$$

W_k^{in} is the rate for scattering to the lower energy polariton modes, i.e. the term responsible for the formation of a BEC. It turns out that this rate is activated for high T_X. For a detailed derivation of this result we refer to [32], and give here only a qualitative explanation: energy-momentum conservation implies that reservoir excitons need an initial energy to jump into the dip shown in Fig. 4. This is possible only for high enough exciton temperatures. Polaritons show, thus, a strange peculiarity: for BEC to be possible, the exciton reservoir must be heated above a threshold temperature, so that W_k^{in} is activated. Note also that one has to consider two different temperatures: the lattice temperature, T_L that can be controlled experimentally and the exciton temperature, T_X.

The set of first-order differential equations (55), must be solved taking into account the dependence of exciton-polariton scattering on T_x, something that must be computed self-consistently. Three mechanisms contribute to determine such temperature:

(i) Heating by the scattering to high energy exciton levels. The total energy is conserved in the exciton-polariton scattering toward the LPB. Thus, the energy lost by the polaritons falling into the lower energy states must be gained by the exciton reservoir.

(ii) Cooling produced by acoustical phonons. The exciton-phonon scattering tries to keep excitons at the lattice temperature, T_L, against the heating considered in *(i)*.

(iii) Cooling to the pump temperature. The injection of excitons into the reservoir by a pump that is considered to be at the lattice temperature tries to keep the exciton reservoir at T_L.

The numerical solution of first-order differential equations (55), together with the equations corresponding to the three contributions to T_X leads to a closed description for the dynamics of the exciton reservoir and lower polaritons.

6.2 Evolution of the Polariton Distribution: Macroscopic Occupation

We study here the polariton distribution created by a non-resonant, continuous pump. On the one hand, this situation is easier to understand than the case of excitation by a laser pulse. In the case of pulsed excitation many excitons decay during the process of relaxation, and the system goes through different exciton and polariton densities. In the case of continuous excitation, the system evolves toward a steady-state where quantities like density and temperature are well defined. This is also the most advantageous situation for polariton BEC, because in the pulsed excitation many polaritons are lost, due to the finite life-time of the cavity photon, before they condense. In order to study the steady-state under continuous pumping, one integrates the set of (55) until the steady solution is reached. The steady solution does not depend, of course, on the initial conditions for the Boltzmann equation.

The most interesting result that one can get from the numerical solution is the calculation of the occupation numbers of the lower polariton modes. Occupation numbers can be experimentally measured, because each mode emits light in a different direction. Angle-resolved measurements allow to study the evolution of the polariton distribution as pump-power (polariton density) is increased. Figure 5 shows a numerical simulation of this distribution for several pump-powers. The polariton distribution has a peak that shifts toward lower energies as pump-power is increased, until it reaches the lowest energy polariton mode, where BEC takes place. It seems rather reasonable that polaritons condense in lower energy modes for large densities. However, in this case, this is a very non-trivial result: it turns out that the shift of the peak in the polariton distribution happens because of the heating of the exciton reservoir. The heating is faster at higher densities, because the exciton-polariton scattering is also faster. As T_X increases, the W_k^{in} rate in (57) is activated, and this condition is what makes possible that excitons jump to the dip in momentum space and, finally, form a Bose–Einstein condensate. Figure 5 shows that the exciton reservoir temperature increases from $T_X \approx 10$ K up to $T_X \approx 25$ K.

When the maximum of the distribution reaches the lowest energy state, very large occupation numbers are achieved. For high pump-power, the main part of the polariton population occupies the lowest energy state. Its occupation is several orders of magnitude larger than that of any other state. This is a requirement for getting polariton condensation. For intermediate pump-powers, occupation numbers larger than one are obtained for modes with $k \neq 0$. It is interesting to estimate the total density (including reservoir excitons and polaritons) n_{XP} at which the relaxation into the ground state is possible. In Fig. 5 pump-powers above 5×10^{10} cm^{-2}/100 ps correspond to polariton densities larger than 4×10^{10} cm^{-2}. Thus, in GaAs microcavities, the weak coupling regime will be reached before ϵ_{max}^{LP} reaches the bottom of the LPB dispersion, due to the small saturation density, and BEC is unlikely.

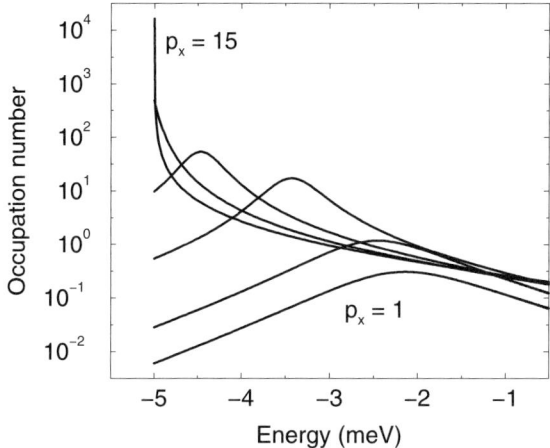

Fig. 5. Evolution of the polariton distribution for $T_L = 10\,\mathrm{K}$, $\Omega_P = 10\,\mathrm{meV}$, and $p_X = 1, 2, 5, 8, 15 \times 10^{10}\,\mathrm{cm}^{-2}/100\,\mathrm{ps}$ (from *bottom* to *top*)

The evolution of the polariton distribution as one increases the pump-power, or the exciton temperature has been demonstrated in an interesting experiment by Savvidis et al. [36]. Experimental results agree very well with the model here described. In such experiment, BEC could not be observed because a GaAs microcavity was used, and the saturation density is reached before the conditions for BEC are met. For the clear observation of the complete evolution of the polariton distribution toward the ground state, CdTe microcavities are much better candidates, because n_{XP} remains well below the saturation density, before the system relaxes toward the ground state.

We focus now on the regime where polaritons can relax into the lowest energy mode. When pump-power is large, and T_X is high enough for the exciton-polariton scattering to be activated, the continuous polariton distribution shown in Fig. 5 saturates, and all the additional polaritons created by the pump fall into the lowest energy mode. This is the BEC regime. The lowest polariton mode corresponds to the bottom of the dip in momentum-space (Fig. 4). If we plot the occupancy in this mode (N_0) as a function of the pump-power, or the polariton density, the transition to very high occupation numbers is very abrupt as shown in Fig. 6, for different CdTe microcavities, and temperatures. Note that the temperatures given in Fig. 6 are the lattice temperatures, but the effective temperature of the exciton reservoir is always higher. The transition occurs always at densities well below n_{sat}, so that the polariton picture is still valid. We stress once more that this true only for CdTe microcavities. In fact, in a CdTe microcavity, experimental evidence of stimulated scattering after non-resonant excitation, that implies $N_0 > 1$, has been demonstrated [37]. In this experiment, it was also observed a narrowing of the linewidth of the light emitted by the polaritons, a property that will be

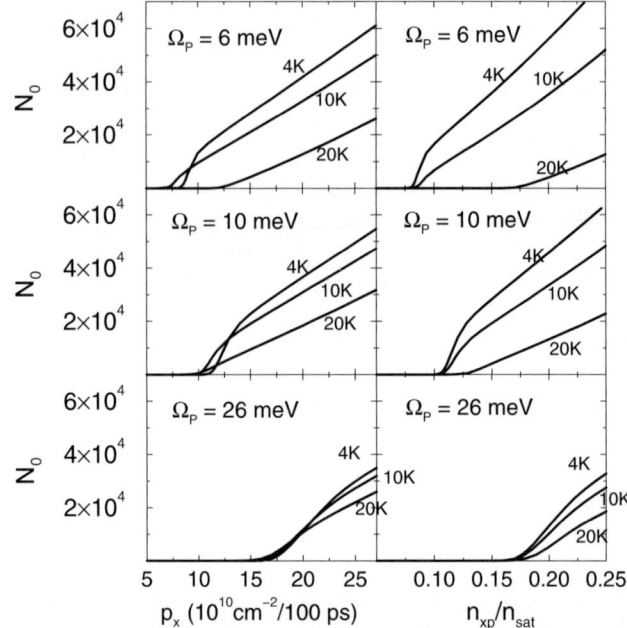

Fig. 6. *Left*: Evolution of the occupancy of the lowest microscopic polariton level, as a function of the pump-power. For each polariton splitting we consider $T_L = 4, 10, 20K$. *Right*: Same as above, but now the occupancy of the lowest microscopic level is plotted as a function of the steady-state exciton-polariton density

discussed in the following section. However, in [37] only the initial stages of the polariton condensation could be achieved because pulsed excitation was used instead of continuous wave excitation. Moreover, polariton lifetime was shorter than the one considered here.

In the case of GaAs microcavities, our calculations show that, unfortunately, the transition to BEC takes place at densities that are comparable to n_{sat}. However, even below the BEC transition, occupation numbers larger than one can be achieved in the polariton modes of a GaAs microcavity. This fact explains why quantum features have been detected in the light emitted by a GaAs microcavity in the experiment of Deng et al. [30].

7 Polariton Laser

The simplest of property of the light emitted by a condensate is the coherence as detected by the narrowing of the linewidth in the spectrum. As already discussed, polaritons are a unique system where the BEC transition, and the emergence of coherence, can be studied in great detail. This is also important from the practical point of view: a BEC of polaritons is expected to emit coherent light, in much the same way as a laser, but without the requirement of population inversion [12]. Our aim in this section is to present a quantitative theoretical description of the spectrum of the emitted light as related to the existence of coherence in polariton condensation.

The standard theory of photon lasers [38, 39], i.e. non-interacting systems, predicts a very narrow linewidth, inversely proportional to the number of condensed particles, $\Gamma^{NI} \propto 1/N_0$. However, polaritons interact with each other through the potential U. This self-interaction in the condensate provokes a process of phase-diffusion that is determined by the energy scale $U N_0$. When $U N_0$ is comparable to Γ^{NI}, this process would increase the linewidth as the number of condensed quasiparticles increases, a behavior that is the opposite to that of a photon laser. A self-consistent framework is required for including these two effects and look for the optimum pumping range for polariton emission.

The intensity of the emitted light can be computed from the first order correlation function (42) which requires the knowledge of the density matrix. Therefore, the steps of the theoretical analysis are: first, to obtain an equation of motion for the density matrix and later to compute correlation functions and emitted intensities.

7.1 Equation of Motion for the Density Matrix

In order to describe the dynamics of the density matrix, we simplify the Hamiltonian to include only the exciton-polariton scattering above discussed and the self-interaction of polaritons in the lowest energy state ($k = 0$) with a macroscopic population. The effective Hamiltonian is the sum of three terms describing the bare-exciton and LP dispersions (H_0), the exciton-polariton scattering (H_{XP}), and the self-interaction in the condensate mode (H_{SI}).

$$H_0 = \sum_{\mathbf{k}} \epsilon_k^{LP} a_{\mathbf{k}}^\dagger a_{\mathbf{k}} + \sum_{\mathbf{k}} \epsilon_X(k) \Psi_{\mathbf{k}}^\dagger \Psi_{\mathbf{k}} ,$$

$$H_{XP} = \sum_{\mathbf{k}} a_{\mathbf{k}}^\dagger F_{\mathbf{k}}^\dagger + h.c.,$$

$$H_{SI} = U a_0^\dagger a_0^\dagger a_0 a_0 . \tag{58}$$

$F_{\mathbf{k}}^\dagger = \sum_{\mathbf{k}_2, \mathbf{k}_3, \mathbf{k}_4} U_{\mathbf{k}, \mathbf{k}_2, \mathbf{k}_3, \mathbf{k}_4} \Psi_{\mathbf{k}_2}^\dagger \Psi_{\mathbf{k}_3} \Psi_{\mathbf{k}_4}$ describes the scattering by means of polariton-polariton interaction U_{k_1, k_2, k_3, k_4} valid for both exciton and LP parts of the spectrum [40]. $U = U_{0,0,0,0}$ is the matrix element for the self-interaction in the lowest energy polariton state. In H_{XP} we have neglected other interactions which only produce energy shifts.

In order to get the quantum dynamics of the condensate, one can trace out the reservoir degrees of freedom in the density matrix operator χ, and define the reduced density matrix s:

$$s(t) = Tr_R\{\chi(t)\}; \qquad \langle O_{LP}(t) \rangle = Tr_{LP}\{O_{LP}(t)s(t)\} \tag{59}$$

where Tr_R, Tr_{LP} represent the trace over the exciton reservoir and the LP, respectively, and O_{LP} is any function of LP operators. The time evolution of the density-matrix is calculated in the interaction picture. Up to the lowest order in $H_{XP} + H_{SI}$, and within a Markovian approximation adequate for the steady state regime, s evolves as given by the master equation:

$$\frac{ds}{dt} = iU\left[s, a_0^{\dagger 2} a_0^2\right] + \sum_k \left(\frac{W_k^{in}}{2}(a_k^{\dagger} s a_k - a_k a_k^{\dagger} s)\right.$$
$$\left. + \frac{W_k^{out} + \Gamma_k}{2}(a_k s a_k^{\dagger} - a_k^{\dagger} a_k s) + h.c.\right) \tag{60}$$

In (60), we have included a term in the Lindbladt form [38] that accounts for the radiative losses with a rate $\Gamma_k = C_k^{LP}/\tau_{ph}$.

Equation (60) allows to calculate the evolution (independent on the self-interaction) of the LP occupation numbers recovering the rate equations (55) that we deduced from the semiclassical Boltzmann equation. On top of the dynamics of the occupation numbers, the master equation (60) allows to study also quantum properties, in particular, the quantum fluctuations that govern the spectrum of the condensate.

7.2 Emission Spectrum

As discussed above, the emission spectrum is obtained by means of (42) from a two-time correlation function which requires the knowledge of the the time evolution of the density matrix we have just discussed. The correlation function for emitted fields $E^{(-)}$ and $E^{(+)}$ is just proportional to the correlation function for polariton operators

$$\langle E^{(-)}(t)E^{(+)}(t+\tau)\rangle \propto \langle a_0^{\dagger}(t+\tau)a_0(t)\rangle \tag{61}$$

which is the magnitude available from $s(t)$.

In order study the dynamics of two time correlation functions, one can use the quantum regression theorem stating that given a set of operators O_j, whose averages satisfy a closed set of linear differential equations:

$$\frac{d}{d\tau}\langle O_j(t+\tau)\rangle = \sum_k L_{j,k}\langle O_k(t+\tau)\rangle, \tag{62}$$

then the two-time averages of O_j with any other operator O, also satisfy the same differential equation:

$$\frac{d}{d\tau}\langle O_j(t+\tau)O(t)\rangle = \sum_k L_{j,k}\langle O_k(t+\tau)O(t)\rangle \tag{63}$$

Without self-interaction in (60), the application of the quantum regression theorem leads to Lorentzian line shape of the spectrum emitted by the condensate mode ($a_0^{\dagger}$):

$$\langle a_0^\dagger(t+\tau)a_0(t)\rangle = N_0 e^{-\Gamma^{NI}\tau}; \Gamma^{NI} = \frac{W_0^{out}+\Gamma_0}{2(1+N_0)}. \tag{64}$$

Decoherence has the usual aspect of a single-particle noise corrected, in the denominator, by the number of particles in the condensate. However, inclusion of the self-interaction term changes this result dramatically. One can use a well-known method in quantum optics that allows to include exactly the effect of the self-interaction [35, 38, 39]. The two-time average is expanded as a sum over a set of auxiliary functions $C_n(t,\tau)$:

$$\langle a_0^\dagger(t+\tau)a_0(t)\rangle = \sum_n \sqrt{n} C_n(t,\tau)$$

$$C_n(t,\tau) = \langle e^{iH_0(t+\tau)}|n\rangle\langle n-1|e^{-iH_0(t+\tau)}a_0(t)\rangle \tag{65}$$

with $|n>$ being the number representation of the $k=0$ state occupation. Using the quantum regression theorem, one can show that the functions C_n satisfy the differential equation [38, 39]:

$$\frac{d}{d\tau}C_n(t,\tau) = i\epsilon_0^{LP}C_n(t,\tau) + \sum_m L_{n,m}C_m(t,\tau) \tag{66}$$

where the $L_{n,m}$ are the coefficients governing the evolution of off-diagonal one-time matrix elements:

$$\frac{d}{d\tau}< n-1|s(\tau)|n> = \sum_m L_{n,m} < m-1|s(\tau)|m>. \tag{67}$$

Using (60) and (67), (66) becomes:

$$\frac{d}{d\tau}C_n(t,\tau) = \left(-w_0^+\,(2n+1) - w_0^-\,(2n-1) + iU(n-1-N_0)\right)C_n(t,\tau)$$

$$+ w_0^+\,2\sqrt{n(n-1)}C_{n-1}(t,\tau) + w_0^-\,2\sqrt{n(n+1)}C_{n+1}(t,\tau) \tag{68}$$

with $w_0^+ = W_0^{in}$ and $w_0^- = (W_0^{out}+\Gamma_0)/2$. In (68) we have taken the origin of energies at $\epsilon_0^{LP}+UN_0$, in order to compare the linewidths of the emission spectrum at different densities. Initial condition for (68) is obtained from C_n for $\tau = 0$:

$$C_n(t,0) = \sqrt{n} < n|s|n> = \sqrt{n}\left(1 - \frac{w_0^+}{w_0^-}\right)\left(\frac{w_0^+}{w_0^-}\right)^n \tag{69}$$

where $< n|s|n >$ is easily deduced from having a time-derivative equal to zero in the steady state. The function $C_n(t,\tau)$ can be interpreted a s the contribution to the spectrum from the state of the condensate mode, with occupation n. When self-interaction is neglected, all these states emit light with the same energy, and decoherence is only provoked by the scattering with non-condensed particles. If N_0 is large enough, then self-interaction cannot be neglected, and now the energy of the emission depends on n. The fact that each possible state in the condensate mode ($k = 0$) emits light with a different energy implies a broadening and a change of the shape of the emission line.

Instead of solving numerically the enormous set of (68), the problem is simplified by replacing the index n by a continuous variable [40, 41]. In the limit $\Gamma^{NI} >> UN_0$, one obtains the adequate Lorentzian shape of the spectrum with a linewidth Γ^{NI}. In the opposite limit, $UN_0 >> \Gamma^{NI}$, an analytic solution exists

$$C(n, \tau) \approx \frac{1}{N_0} \sqrt{n} e^{-\frac{n}{N_0} + iU(n-N_0)\tau} \tag{70}$$

predicting an asymmetrical lineshape:

$$I(\epsilon) \approx \frac{1}{U} \frac{(\epsilon + UN_0)}{UN_0} e^{-(\epsilon+UN_0)/(UN_0)} \theta(\epsilon + UN_0) . \tag{71}$$

Let us discuss the emission spectrum in the particular case, once again, of a CdTe microcavity going from the non-interacting limit to the strongly interacting one through an intermediate regime which requires a numerical calculation [40]. We take $\Omega_P = 10$ meV, and zero detuning between the bare exciton and the photonic mode. The pump is assumed to add excitons at a lattice temperature, $T_L = 10$ K. Lifetimes of the photon and the bare excitons are taken $\tau_{ph} = 1$ ps and $\tau_X = 100$ ps. The steady-state polariton density we considered here is always below 0.3 times the saturation density. Figure 7 gives N_0 as a function of the pump-power. It shows a threshold for BEC ($N_0 > 1$) around $p_X \approx 810^8$ cm^{-2} ps^{-1}. For densities below the condition $N_0 U \approx \Gamma^{NI}$, the calculated lineshape is Lorentzian with optimal conditions giving a minimum linewidth of the order of 1 μeV. When $N_0 U \approx \Gamma^{NI}$ is reached, there is a transition from the laser-like decoherence to the self-interaction broadening as shown in Fig. 8. Increasing the pump-power, produces an asymmetric emission with a shape given by (71).

In the evolution of the spectrum linewidth as a function of the pump-power, shown in Fig. 7, one observes that the optimum pump is very well defined by the abrupt dip in the emission linewidth. This dip is due to an increase of the coherence of almost three orders of magnitude. For larger densities the linewidth of the polariton laser increases linearly with the number of condensed particles, until it reaches values comparable to the non-condensed emission. This feature is a very important conclusion: polariton lasers show self-interaction effects that must be properly taken into account in order to predict the final line-width of the emitted light. This conclusion corrects the usual picture that we had about exciton and polariton BEC. The experimental detection of all these features would be a direct determination of the lasing of polaritons in the microcavity.

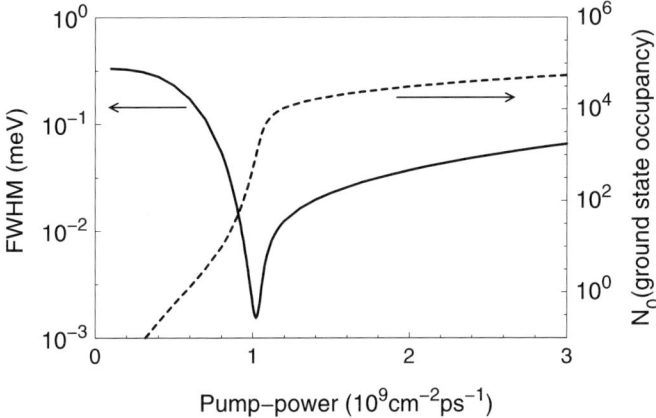

Fig. 7. Full Width at Half Maximum of the emission spectra (*continuous line*) and ground state occupancy (*ashed line*) as a function of the pump-power p_X (in $10^9 \, \mathrm{cm}^{-2} \, \mathrm{ps}^{-1}$). The optimum pump corresponds to $p_X \approx 10^9 \, \mathrm{cm}^{-2} \, \mathrm{ps}^{-1}$

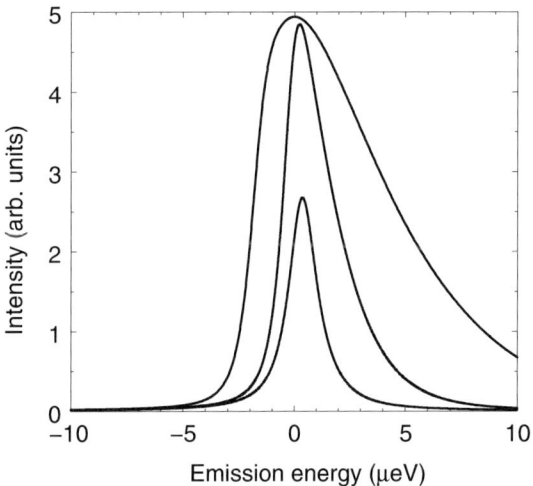

Fig. 8. Transition from non-interacting coherence, to self-interaction decoherence in the emission spectra for $p_X = 1.02, 1.05, 1.10$ in units of $10^9 \, \mathrm{cm}^{-2} \, \mathrm{ps}^{-1}$

7.3 Related Work

We finish this section about polariton condensation with a brief outlook to other works on this topic. Polariton condensation has attracted a lot of attention, both from theorists and experimentalists. We mention here a few research lines that seem to be very promising in the future of this field.

Experimentalists are looking for materials where excitons and polaritons are more stable, and survive up to higher densities. Good candidates seem to be ZnO [42], and GaN based microcavities, where the exciton binding energy is even larger than in the case of CdTe. Many of our results about polariton dynamics would also apply to this case, for example, the activation of the exciton-polariton rate with exciton temperature, and the evolution of the polariton distribution as one increases the pump-power.

A different approach is to consider relaxation mechanisms different from the exciton-polariton scattering considered here, in order to shorten the formation of the polariton BEC. If semiconductor microcavities are doped, there exists a population of free electrons that can provide an efficient relaxation mechanism for the formation of the polariton condensate. This idea, proposed by Malpuech et al. [43], has been studied in recent experiments [44, 45]. If a condensate is formed by this mechanism, the emitted light would show features similar to those presented in the previous subsection, because the narrowing of the linewidth, and the effect of self-interaction do not depend on the particular scattering mechanism that is responsible for the formation of the BEC.

Finally, we point out that, as discussed for excitons, the spin degree of freedom of a polariton condensate opens many exciting research lines. Polaritons exist in two different polarizations, that correspond to those of their excitonic and photonic components. The measurement of the degree of polarization of the emitted light allows to study how the effective polariton spin evolves during the process of thermalization and condensation. Recent experiments [46] on the spin dynamics of non-resonantly excited polaritons have shown that the polarization of the light emitted by the polaritons presents either an enhancement or a reversal depending on the detuning of the exciton with the cavity photon. This effect can by explained by the effect of stimulated scattering of polaritons during the relaxation.

8 Summary

In these notes, we present a theoretical description of a condensate of either excitons or exciton-polaritons in semiconductors. This text is not exhaustive on all the aspects of the field. In particular, we do not include a description on the experimental situation which is currently developing very fast [1]. Moreover, excellent descriptions of many topics in this field are contained in references [2, 3].

We start by presenting the standard theory. We do not study the dynamics of condensation but instead, we concentrate in the question of detecting condensation. This is essentially made by studying the emission of light from the exciton condensate. The case of condensation of excitons in the presence of a high magnetic field is briefly discussed. The third component of the total angular momentum of an exciton represents an internal degree of freedom opening the extremely interesting possibility of multicomponent condensates. We discuss this possibility in some detail and find a possible transition between polarized and non-polarized condensates. Moreover, we discuss the interesting behavior of the polarization of the light emitted by this multicomponent condensate.

An important problem for this type of condensation is the exciton lifetime and its large effective mass which implies a large threshold density for condensation. A possible alternative is to consider exciton-polaritons in quantum wells embedded in microcavities. In this case we carefully describe the dynamics (stimulated scattering) driving to a large density of these quasiparticles in a given state, a necessary ingredient for condensation. This large occupation allows the existence of a polariton laser which shows special characteristics in its spectrum.

As a final comment, we must stress that we follow the terminology used in the field although the term *"condensation"* is not the most adequate because we are working with a system out of equilibrium. In fact, our system is closer to a laser and this similarity is responsible for the current large activity in this interesting field. In particular, the main theoretical challenge is, probably, the understanding of the dynamics driving to the condensation. This is a topic that will receive much attention in the near future.

Acknowledgments

Work supported in part by MCYT of Spain under contract No. MAT2002-00139, CAM under Contract No. GR/MAT/0099/2004 and UE within the Research Training Network COLLECT.

References

1. D. Snoke, Science, **298**, 1368 (2003) and references therein.
2. S.A. Moskalenko and D.W. Snoke, *Bose-Einstein condensation of excitons and biexcitons*, (Cambridge University Press, Cambridge, 2000)
3. *Bose-Einstein condensation*, A. Griffin, D.W. Snoke and S. Stringari, eds. (Cambridge University Press, Cambridge, 1995)
4. H. Haug and A.P. Jauho, *Quantum kinetics in transport and optics of semiconductors*, (Springer, Berlin, 1996)
5. F. Rossi and T. Kuhn, Rev. Mod. Phys., **74**, 895 (2002) and references therein.
6. Y. Yamamoto and A. Imamoglu, *Mesoscopic quantum optics*, (Wiley, New York, 1999)
7. P. Nozieres in ref. [3].
8. L.V. Keldysh and Yu.V. Kopaev, Fiz. Tverd. Tela **6**, 2791 (1964) [Sov. Phys. Solid State **6**, 2219 (1965)]; L.V. Keldysh and A.N. Kozlov, Sov. Phys. JETP **27**, 521 (1968)
9. C. Comte and P. Nozières, J. Physique **43**, 1069 (1982); P. Nozières and C. Comte, *ibid* **43**, 1083 (1982)
10. H. Haug and S. Schmitt-Rink, Prog. Quant. Electr., **9**, 3 (1984)

11. L.V. Keldysh, in ref. [3]
12. A. Imamoglu, R.J. Ram, S. Pau and Y. Yamamoto, Phys. Rev. A **53**, 4250 (1996)
13. J. Fernandez-Rossier, C. Tejedor and R. Merlin, Solid St. Commun., **108**, 473 (1998)
14. H. Haug and S.W. Koch, *Quantum Theory of Optical and Electronic Properties of Semiconductors* (World Scientific, London, 1993)
15. P.W. Anderson, *Basic Notions of Condensed Matter Physics* (Benjamin-Cummings, Menlo Park, 1984)
16. D. Forster, *Hydrodynamic Fluctuations, Broken Symmetry, and Correlation Functions* (Benjamin, New York, 1975)
17. S. Stringari in ref. [3]
18. X. Zhu, P.B. Littlewood, M.S. Hybertsen and T.M. Rice, Phys. Rev. Lett. **74**, 1633 (1995); P.B. Littlewood and X. Zhu, Physica Scripta **T68**, 56 (1996)
19. A. Olaya-Castro, F.J. Rodríguez, L. Quiroga, and C. Tejedor, Phys. Rev. Lett. **87**, 246403 (2001)
20. L.V. Butov, C.W. Lai, A.L. Ivanov, A.C. Gossard and D.S. Chemla, Nature **417**, 47 (2002); L.V. Butov, A.C. Gossard and D.S. Chemla, Nature **418**, 751 (2002)
21. D. Snoke, S. Denev, Y. Liu, L. Pfeiffer and K. West, Nature **418**, 754 (2002)
22. D. Paquet, T.M. Rice and K. Ueda, Phys. Rev. B **32**, 5298 (1985)
23. S.M. Girvin and A.H. MacDonald in *Perspectives in quantum Hall effects*, edited by S. DasSarma and A. Pinczuk (Wiley, New York, 1997)
24. A.J. Leggett, *Rev. Mod. Phys.* **47**, 331 (1975)
25. J. Fernández-Rossier and C. Tejedor, Phys. Rev. Lett. **78**, 4809 (1997)
26. L. Viña, L. Muoz, E. Prez, J. Fernndez-Rossier, C. Tejedor and K. Ploog, *Phys. Rev. B* **54**, R8317 (1996)
27. M.Z. Maialle et al., *Phys. Rev. B* **47**, 15776 (1993)
28. A.J. Leggett and F. Sols, Found. Phys. **21**, 353 (1991)
29. P.R. Eastham and P.B. Littlewood, Phys. Rev. B **64**, 235101 (2001); M.H. Szymanska, P.B. Littlewood and B.D. Simons, Phys. Rev. A **68**, 013818 (2003)
30. H. Deng, G. Weihs, C. Santori, J. Bloch and Y. Yamamoto, Science, **298**, 199 (2002)
31. S. Schmitt-Rink, D.S. Chemla, and D.A.B. Miller, Phys. Rev. B **32**, 6601 (1985)
32. D. Porras, C. Ciuti, J.J. Baumberg and C. Tejedor, Phys. Rev. B **66**, 085304 (2002)
33. C. Piermarocchi, F. Tassone, V. Savona, A. Quattropani, and P. Schwendi-mann, Phys. Rev. B **55**, 1333 (1997)
34. F. Tassone and Y. Yamamoto, Phys. Rev. B **59**, 10830 (1999)
35. M. Holland, K. Burnett, C. Gardiner, J.I. Cirac and P. Zoller, Phys. Rev. A **54**, R1757 (1996)
36. P.G. Savvidis, J.J. Baumberg, D. Porras, D.M. Whittaker, M.S. Skolnick, and J.S. Roberts, Phys. Rev. B **65**, 073309 (2002)
37. Le Si Dang, D. Heger, R. André, F. Bœuf, and R. Romestain, Phys. Rev. Lett. **81**, 3920 (1998)
38. D.F. Walls and G.J. Milburn, *Quantum optics*, (Springer-Verlag, Berlin, 1994)
39. M.O. Scully and M.S. Zubairy, *Quantum optics*, (Cambridge University Press, Cambridge, 1997)
40. D. Porras and C. Tejedor, Phys. Rev. B **67**, 161310(R) (2003)

41. G.W. Gardiner *Handbook of stochastic methods*, Springer, Berlin (1996)
42. M. Zamfirescu, A. Kavokin, B. Gil, G. Malpuech, and M. Kaliteevski, Phys. Rev. B **65**, 161205 (2002)
43. G. Malpuech, A. Kavokin, A. Di Carlo, J.J. Baumberg, Phys. Rev. B **65**, 153310 (2002)
44. P.G. Lagoudakis, M.D. Martín, J.J. Baumberg, A. Qarry, E. Cohen, and L.N. Pfeiffer, Phys. Rev. Lett. **90**, 206401 (2003)
45. A.I. Tartakovskii, D.N. Krizhanovskii, G. Malpuech, M. Emam-Ismail, A.V. Chernenko, A.V. Kavokin, V.D. Kulakovskii, M.S. Skolnick and J.S. Roberts, Phys. Rev. B **67**, 165302 (2003)
46. M.D. Martín, G. Aichmayr, and L. Viña, Phys. Rev. Lett. **89**, 077402 (2002)

Lecture Notes in Physics

For information about earlier volumes
please contact your bookseller or Springer
LNP Online archive: springerlink.com

Printing: Krips bv, Meppel
Binding: Stürtz, Würzburg